ELECTRICAL INSTALLATION

REVISION WORK

Part II Level

Author

E. G. Stocks

Published by CT Projects

Published in Great Britain 1992

by CT Projects

Depot Road

HORSHAM

ISBN 1 871230 43 8

Table of Contents

Introduction

Notes about this book:

This book has been specifically designed to help with revision for courses such as the City and Guilds Electrical Installation Competences Course 236 Part II, the City and Guilds Electrical Installation Technology Course 8230 Part II (overseas) and appropriate SCOTVEC and BTEC courses.

This is not a text book or a book of course notes.

Five subject areas have been identified and are covered in Sections Two to Six of this book. The layout of each section is similar and at the front of each a number of questions have been set on similar lines to those found in examination papers. Each section has then been divided up so that the subject area around each question is dealt with separately. The questions have been designed to cover different parts within specific subject areas. Before an answer is given to any question the theory associated with it is explained. This gives a greater depth and helps when having to answer similar questions on the same subject area.

To help to pick out the important "key points" when scanning back over the text the symbol has been used.

An example answer is given to each question showing the minimum that should be included. These answers have been put together on the basis of gaining a pass in an examination.

So that the revision can be taken further a question has been set under "Now try this". Help in answering these can be found at the end of each section. The exact answers are not shown but "Tips" are included so as to give guidance as to where the answers may be found. More revision questions on each subject area have been set at the end of each section.

To give practice in answering multiple-choice questions a 50 item paper is included at the back of this book.

Although a revision book cannot cover an entire syllabus every effort has been made to cover the important points especially where there have been common errors or misconceptions in the past.

Section One

Notes on Revision

Why do you revise?

The decision as to whether or not to revise is in the end up to each individual. Some people will argue that there should be no need to revise, others think it will make up for a lack of study at earlier stages. Although everybody is different and has their own requirements, their ultimate aim is usually the same and that is to pass the examination.

It is often quite a long time from the start of a course of study to the time of the examination. In that time new subjects have been introduced and details or original points forgotten. Some areas of study may also have been misunderstood. Revision should be seen as a way of bringing subjects back to mind and putting them in some order.

Revision should not be seen as a substitute for original study but should be used to go over work covered earlier and to fill in any gaps.

It is always better to turn up on the day of the examination feeling confident, and good, well-planned revision can help.

When do you revise?

In most cases the night before the examination is too late to start revising. Similarly, before the subject has been covered is too early. Almost anytime in between may be suitable.

As everybody is different it is not possible to say what day of the week or what time of day is best to revise. There are, however, some points that should be considered. It takes time to revise properly and that time has to be found and set aside. Revision also requires concentration and this usually means finding a quiet place where there will be little or no interference. Ideally, the revision should be carried out when the mind is fresh and responsive, not late at night when it is fighting to stay awake.

How do you revise?

Before starting to revise it is important to draw up a programme of work. To help to determine which subject areas are required the questions at the front of each section should be attempted without referring to the rest of the section. This exercise will also give an indication as to the depth that each subject area needs revising to.

A good revision programme should include the subject area to be covered, the type of questions that should be answered and a target time each part should take to complete.

The following guide may help when planning a revision session using this book.

- Select the subject area to revise
  - > go to that section in the book

- Read through the questions at the front of the section and select the one to revise
  - > try the question on rough paper
  - > go to that area in the section

- Read through the background theory and the answer to the question
  - > check your answer for inaccuracies and/or omissions
  - > look up information in other books if necessary
  - > make notes to remember facts

- Attempt the "Now try this" question
  - > try to answer the question on rough paper without referring to the rest of the section
  - > look up the "Tips" at the end of the section
  - > go back over the "keypoints"
  - > look up information in other books if necessary
  - > complete the answer in the space provided

- Select another similar question from the end of the section
  - > look up details
  - > answer the question

As a check on what needs further revision try the 50 multiple-choice questions at the back of the book.

The Examination

There are a number of other points that should be considered when revising that can help when taking the examination. Where books such as the "IEE Wiring Regulations" and "On Site Guide" can be used in the examination room these need to be studied when revising so that the relevant information can be found quickly. Even with these publications available there are going to be some facts that have to be remembered. It can be helpful to build a story up around them, fit them into a rhyme or visualise them in picture form. This applies equally to formulae and mathematical equations.

One final point - good clear presentation with labelled diagrams could make just the difference between a pass and a failure!

Section Two

Supply Systems

This section will look at these 6 questions and the subject area around them.

1.a. A delta/star transformer supplies a factory at 415V three-phase and 240V single-phase. Draw a labelled diagram to show how the following can be obtained from the star connected transformer winding.

    i)   415V, three-phase 3 wire
    ii) 415V, three-phase 4 wire
    iii) 240V, single-phase
    iv) 415V, single-phase

  b. Calculate the line current in each phase if a balanced load of 75kW at a power factor of 0.75 lagging is supplied with 415V three-phase.

2. The 240V supply to a small market garden is by 2-single phase overhead cables. Explain:

    a. how the consumer's installation should be protected from possible earth faults
    b. the difference between the earthing arrangements of this and that of a TN-C-S system.

3. The electrical equipment at a mains intake to a factory consists of:

    i)   500A TPN fused switch
    ii) 500A TPN busbar chamber
    iii) 100A TPN switch fuse supplying a 4 way TPN distribution fuse board at the intake position
    iv) 60A DP switch fuse supplying a single-phase 4 way consumer unit some distance from the intake position
    v) 100A TP switch fuse supplying a 6 way TP distribution fuse board some distance from the intake position

Draw and label a circuit diagram showing the connections of each of the above. It is not necessary to show the Electricity Company equipment.

4. A number of machines are to be supplied from an overhead busbar trunking system.

    a) Explain with the aid of a diagram how the floor mounted machines can be connected to the busbar trunking.
    b) State two advantages of an overhead busbar system compared with a rewirable trunking and conduit system.

5. Three resistive single-phase loads are connected to a three-phase distribution board. The loads are:
    Red phase 12A
    Blue Phase 5A
    Yellow phase 7A

Using a scale of 10mm = 1A construct a phasor diagram and measure the current flowing in the neutral of the supply to the distribution board.

6. The components shown in Diagram 2.1 are those required to produce a simple battery charger. Redraw these to create a circuit suitable for a battery charger unit. Show all d.c. polarities and label each component.

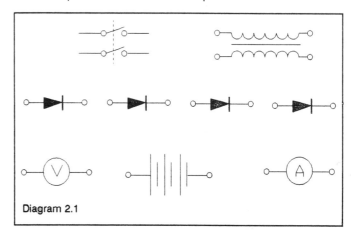

Diagram 2.1

Introduction

It is important to remember that in this section the subject area covers the supply of electricity from the substation at 11kV down to the consumer and throughout the installation. It can also apply to the supply from sources such as batteries and power supply units. This gives a broad area of study and there are many questions that can be asked.

Three-phase Supplies

It is quite common to find a factory with a substation transformer adjacent to it, or even built into the premises. The supply to the transformer is usually 11kV to a delta connected winding. This means that only three conductors are connected to this side of the transformer. The output is normally from a star connected winding delivering the voltages required within the factory. These voltages can be categorised under two headings, namely single-phase and three-phase.

Single-phase

Single-phase supplies are usually associated with one phase and the neutral star point of the transformer windings. However, this is not always the case for single-phase supplies can be obtained by connecting to any of the two phases. The voltages are of course different in each case. The supply from any one phase and neutral will normally be 240V, as shown in Diagram 2.2, whereas connections between any two phases will give 415V. Diagram 2.3

Three-phase

A three-phase supply is obtained, as the name implies, from all three phases. This normally gives a voltage of 415V between any of the phases. Diagram 2.4

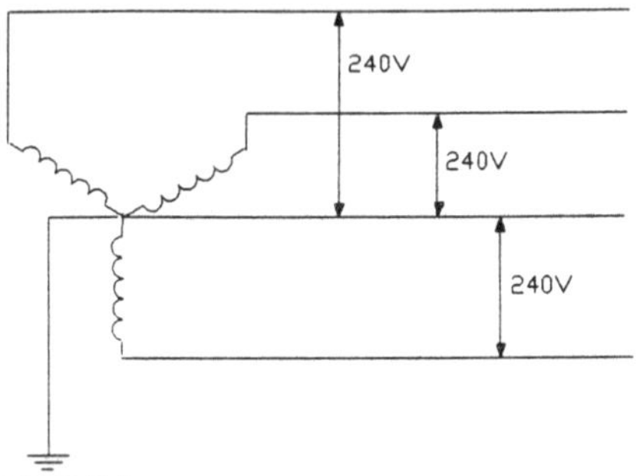

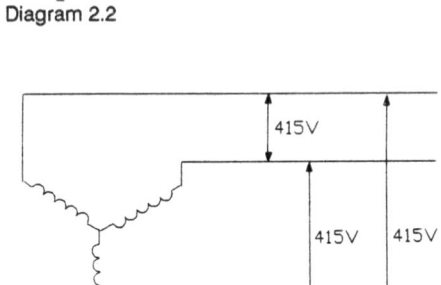

240V
240V
240V

Diagram 2.2

415V
415V 415V

Diagram 2.3

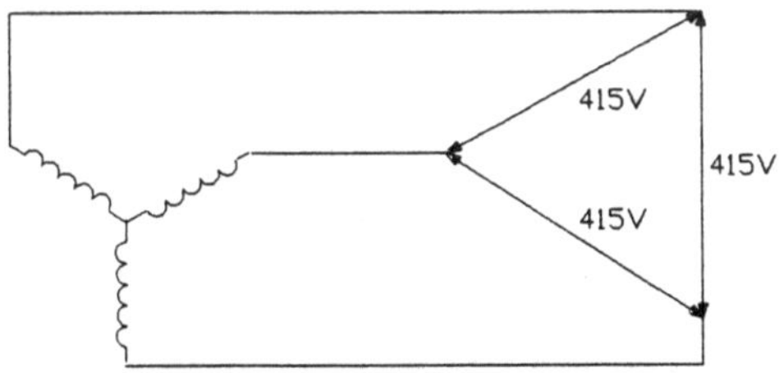

415V
415V
415V

Diagram 2.4

Power

The power dissipated by a balanced three-phase load can be calculated from

$$P = UI \sqrt{3} \cos \varnothing$$

where
P = Power dissipated
U = Voltage between phases
I = Current in each line
cos Ø = Power factor

From this formula any of the other values can be calculated by transposing.

So where the output winding of a transformer is connected as a star with the centre point common to all three-phases, a combination of voltages can be obtained. These consist of 240V and 415V single-phase and 415V three-phase.

Question 1

1.a. A delta/star transformer supplies a factory at 415V three-phase and 240V single-phase. Draw a labelled diagram to show how the following can be obtained from the star connected transformer winding.

  i) 415V, three-phase 3 wire
  ii) 415V, three-phase 4 wire
  iii) 240V, single-phase
  iv) 415V, single-phase

b. Calculate the line current in each phase if a balanced load of 75kW at a power factor of 0.75 lagging is supplied with 415V three-phase.

Answer

1a) The supplies obtained from a delta star transformer.

Diagram 2.5

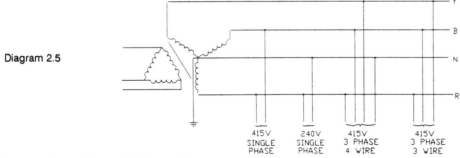

415V SINGLE PHASE	240V SINGLE PHASE	415V 3 PHASE 4 WIRE	415V 3 PHASE 3 WIRE

b) Power in three-phase (P) = UI √3 cos Ø

  Transpose for I

$$I = \frac{P}{U \sqrt{3} \cos Ø} = \frac{75 \times 1000}{415 \times 1.73 \times 0.75}$$

$$= 139.3A$$

Now try this - One

a. Four single-phase loads are to be connected to a three-phase 415-240V star connected transformer. Three of these are to supply 240V loads but the fourth is to supply a 415V load. Draw a diagram to show how the voltages for each load can be obtained.

b. Calculate the power, in kilowatts, of one 240V single phase load if a current of 22A at a power factor of 0.85 is flowing.

Supply Systems

When a supply company connects the electricity supply to a consumer's premises they will normally not only be able to give phase and neutral but also an earth connection. The supply substation transformer automatically has to have a connection to earth on either the star point of a three-phase supply or one side of a single-phase supply. This connection is not only to earth but also to the neutral. Diagrams 2.6 and 2.7

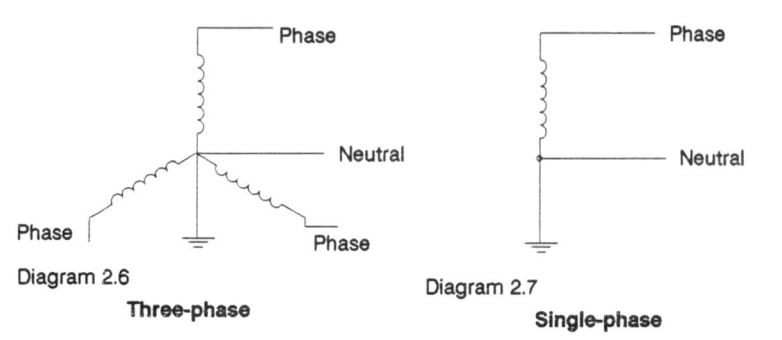

Diagram 2.6
Three-phase

Diagram 2.7
Single-phase

The earth from the substation to the consumer may be a separate  conductor in the supply cable or one that is combined with the neutral conductor. When it is separate this is usually in the form of the steel wire armouring around the cable, Diagram 2.8. Often now it is combined and is in the form of a "concentric" cable where the neutral/earth conductor surrounds the phase conductor. Diagram 2.9

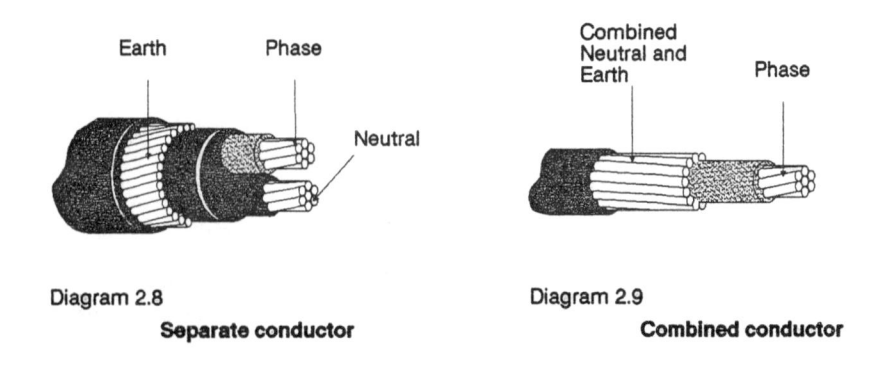

Diagram 2.8
Separate conductor

Diagram 2.9
Combined conductor

The supply company is under an obligation to notify the consumer of the method of earthing that is used. This is included in the installation's supply system details. For example in a TN-S system the "S" denotes that the earth connection to the consumer is by a separate conductor. In the TN-C-S system however, the "C" shows that the earth conductor is combined with the neutral for the supply but in the installation it must be separate.

There are occasions where the supply company is unable to give an earth connection at the consumer's premises. These are usually in situations where the supply to the premises is by overhead cables and only the phase and neutral conductors are supplied. These supplies are referred to as TT systems. In these situations consumers are responsible for their own earth protection arrangements. These usually consist of a combination of a consumer's earth electrode and a residual current device.

Remember that under fault conditions either enough current should flow to make the protective device operate in the required time, or, where this may not be possible an rcd is used to detect faults to earth before they can become dangerous. It is seldom possible to use just an earth electrode, especially if the installation includes socket outlets or loads protected by devices rated at 30A or above. It therefore becomes standard practice to provide an rcd and earth electrode for TT systems. Diagram 2.10

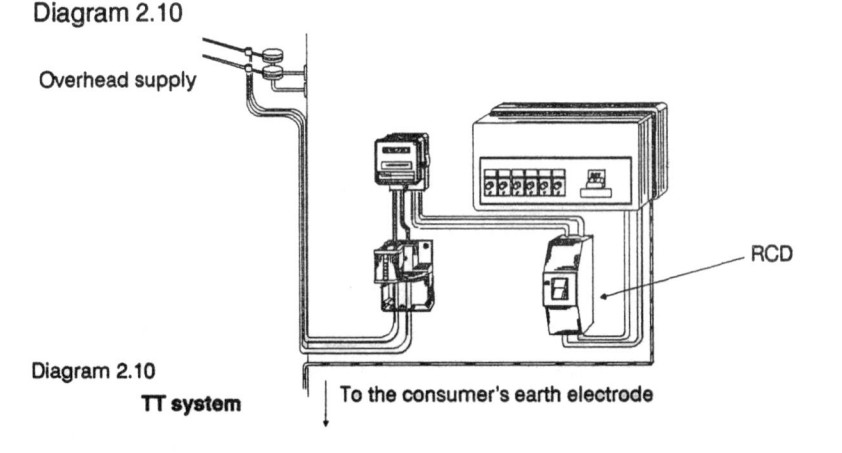

Diagram 2.10
TT system

To the consumer's earth electrode

Question 2

The 240V supply to a small market garden is by 2 single phase overhead cables.
Explain:

   a. *how the consumer's installation should be protected from possible earth faults*
   b. *the difference between the earthing arrangements of this and that of a TN-C-S system.*

Answer

a. As the supply to the market garden is by two cables only the supply company does not include a conection to earth. The supply is in effect a TT system and the consumer has to provide the earth fault protection. This will consist of an rcd and an earth electrode in addition to the overcurrent protection equipment.

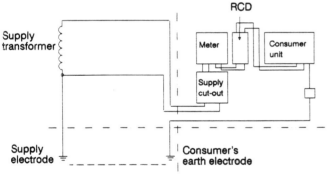

Diagram 2.11

TT System

Under fault conditions the current flows through the phase conductor but returns through the protective conductor not the neutral. The rcd detects the phase and neutral currents are not the same and automatically switches off the electrical supply.

b. As we have seen in (a) the supply company does not include a connection to earth with a TT system. A TN-C-S system is supplied with a cable where the earth connection and the neutral conductor are combined. At the consumer's premises these are separated so that as far as the consumer is concerned it is a phase, neutral and earth supply. It is not essential to include an rcd with all TN-C-S systems but they are sometimes added for extra protection.

Now try this - Two

Draw the sequence of control equipment in a consumer's premises for a TN-S system. Include and label the supply company's and consumer's equipment.

Intake Distribution

Once the supply company has identified the supply it will provide, it is the consumer's responsibility to ensure that their installation meets the necessary requirements.

To avoid danger in the event of a fault an installation is divided up into a number of separate circuits.

To ensure safety all circuits must be designed to include:

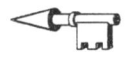

- overcurrent protection
- means of isolation
- adequate means of earth fault protection

Should a fault occur it is essential that only the circuit with the fault is automatically disconnected. Protection devices must therefore be graded so that the one closest to the fault will operate and others will not be affected. This is known as discrimination. Diagram 2.12

On a single-phase domestic type intake this is comparatively simple for each circuit has its own protection device within a consumer unit. The means of isolation is the main switch and the earth fault protection is either through the protection devices or a separate rcd. Diagrams 2.13 and 2.14

Where a number of loads have to be taken to separate circuits at the main intake, such as on a three-phase supply, a busbar chamber is often used as a large junction box. This chamber, as with anywhere else on the installation, must have means of over-current protection and isolation. To ensure this a switch fuse, or fused switch, is connected on the supply side. Each load taken from the busbar chamber is first connected to a switch fuse to ensure its initial protection. Diagrams 2.15 and 2.16

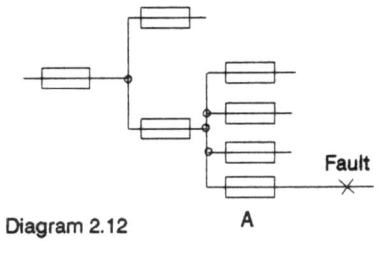

Diagram 2.12

Only fuse A operates and none of the others are affected.

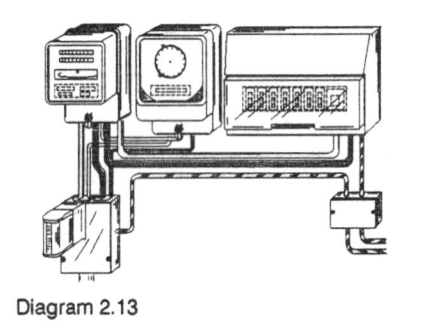

Diagram 2.13

Single-phase distribution system

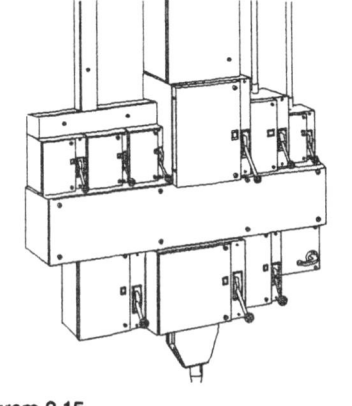

Diagram 2.15

Three-phase distribution system

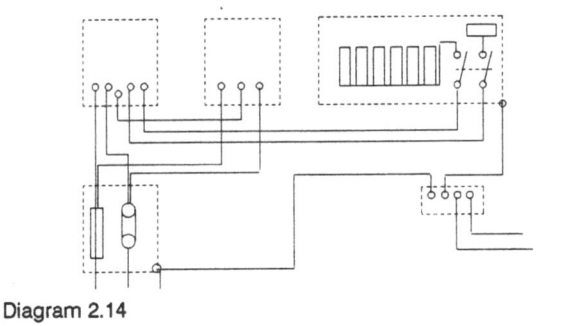

In some cases the load or loads are some distance away from the intake position. Where this is the case separate means of isolation will be necessary at the load or distribution board.

Solid metal connections must be made throughout to ensure a good earth fault path is always available.

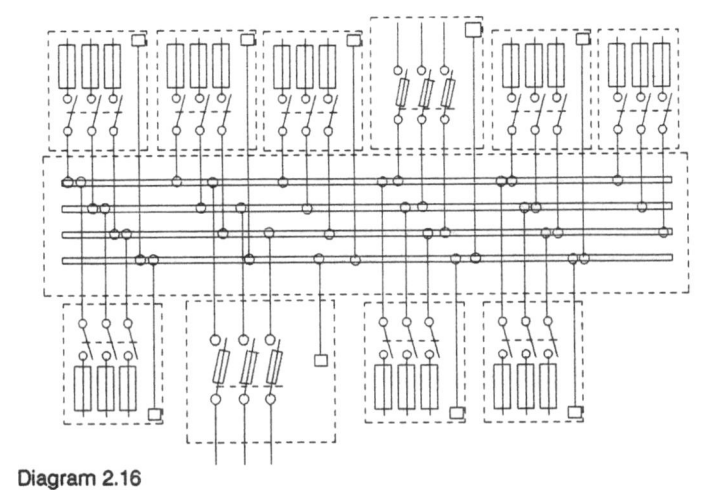

Diagram 2.14

Circuit diagram for the single-phase distribution system.

Diagram 2.16

Circuit diagram for the three-phase distribution system.

Question 3

The electrical equipment at a mains intake to a factory consists of:

i) 500A TPN fused switch
ii) 500A TPN busbar chamber
iii) 100A TPN switch fuse supplying a 4 way TPN distribution fuse board at the intake position
iv) 60A DP switch fuse supplying a single-phase 4 way consumer unit some distance from the intake position
v) 100A TP switch fuse supplying a 6 way TP distribution fuse board some distance from the intake position

Draw and label a circuit diagram showing the connections of each of the above.
It is not necessary to show the Electricity Company equipment.

Answer

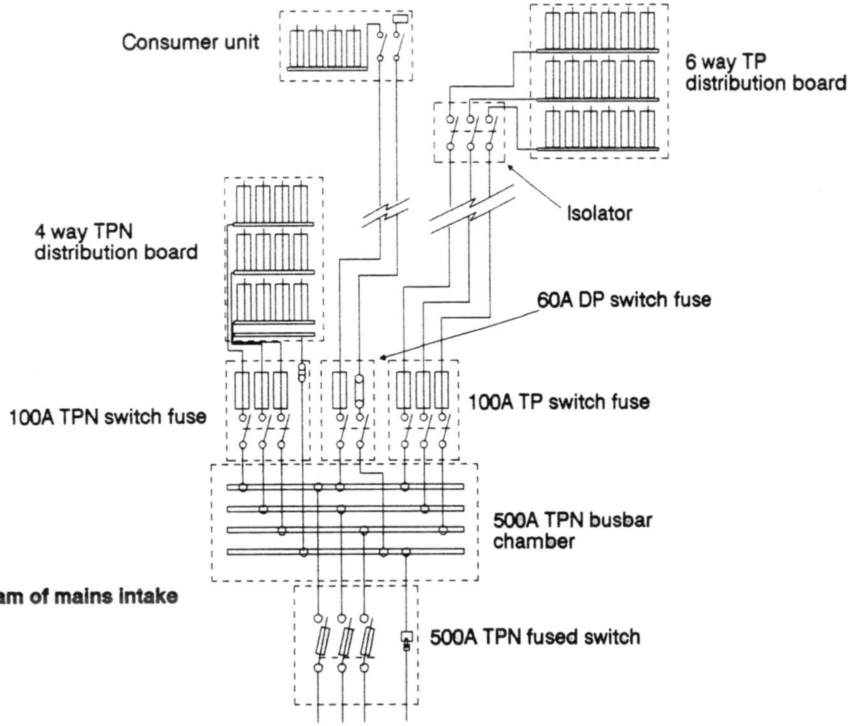

Diagram 2.17

Circuit diagram of mains intake to a factory.

Labels in diagram: Consumer unit, 6 way TP distribution board, Isolator, 4 way TPN distribution board, 60A DP switch fuse, 100A TPN switch fuse, 100A TP switch fuse, 500A TPN busbar chamber, 500A TPN fused switch

Explain what is meant by discrimination when applied to circuit protection. Draw a simple line diagram to show discrimination is allowed for on a load plugged into a BS 1361 socket on a domestic ring final circuit.

Internal Distribution

The distribution of electrical supplies within a building can be carried out in several different ways. The particular method adopted may depend on many different factors. In a factory where supplies have to be connected to machines there are special considerations that must be given. For a start, the environment is often such that mechanical damage to electrical equipment must be considered. Machines are sometimes moved to different positions. The fact that many machines vibrate due to moving parts means that electrical connections must allow for this.

Systems using a mixture of steel trunking and conduit have often been used in factory installations. These can usually meet the requirements with regard to mechanical damage. They do not however always give the flexibility required when alterations have to be made.

In more recent years overhead busbar trunking has become an accepted method of supplying equipment within the factory environment. The steel trunking has solid copper, or aluminium, rods or bars fitted throughout its length, mounted on insulators. At regular points there are positions that allow for tapping off to loads. These tap off points are usually plug-in fuse or mcb units so that local overcurrent protection and isolation is automatically achieved. When a tap off point is not in use there are no exposed conductors as shutters automatically come across covering the busbars. Diagram 2.18

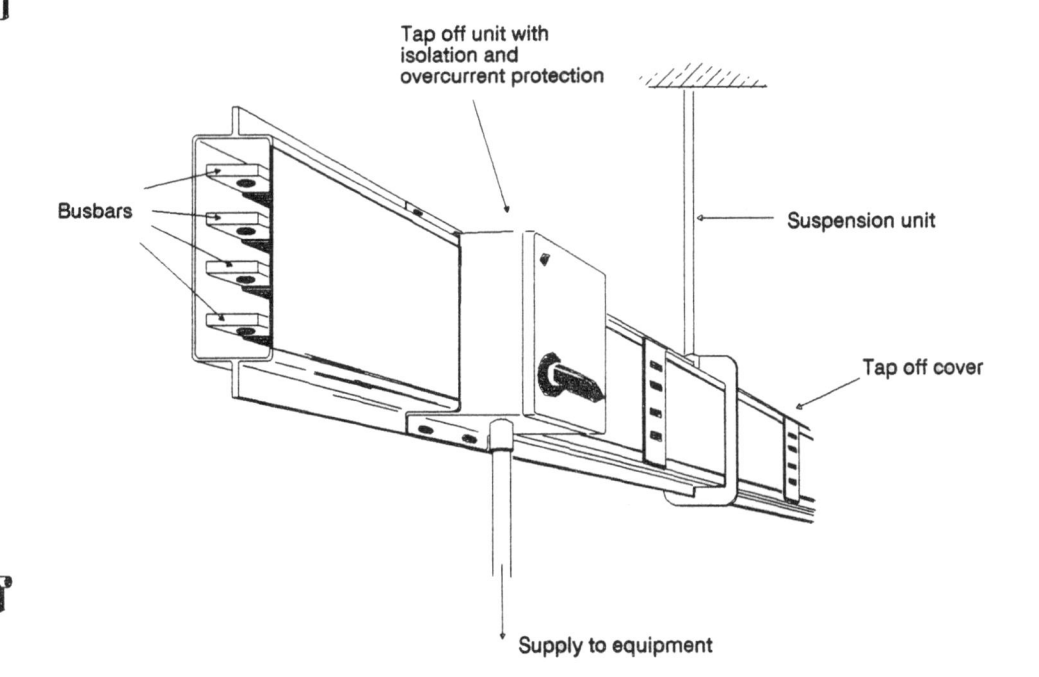

Diagram 2.18

Overhead busbar system

Question 4

A number of machines are to be supplied from an overhead busbar trunking system.

a. *Explain with the aid of a diagram how the floor mounted machines can be connected to the busbar trunking.*
b. *State two advantages of an overhead busbar system compared with a rewirable trunking and conduit system.*

Answer

a. A floor mounted machine can be supplied from an overhead busbar trunking system using manufactured tap off points. These points provide means of overcurrent protection and may also incorporate an isolator. The method of dropping from the tap off point to the machine may depend on their relative positions. Assuming that there is a tap off point above the machine and this is away from any structural part of the building, a steel conduit drop can be used as shown in the diagram.

This method gives mechanical protection and allows for vibration in the machine.

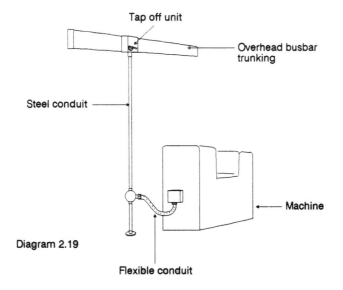

Diagram 2.19

b. Two advantages of an overhead busbar system compared with a rewirable trunking and conduit system are:

i) It is comparatively simple to move the supply to the machine to a new position without carrying out major rewiring.

ii) The overcurrent protection and isolation is part of the system and does not have to be provided separately .

(There is also the advantage that voltage drop does not become a major problem when moving machines further away from the intake position.)

Now try this - Four

Explain why a steel conduit drop from an overhead trunking can be preferable to trailing flexible cables.

Three-phase Loads and Neutral Currents

Whenever three phases are used to supply different loads there is the problem of trying to keep them in balance. The greater the out of balance of the phases the greater the current that flows in the neutral conductor.

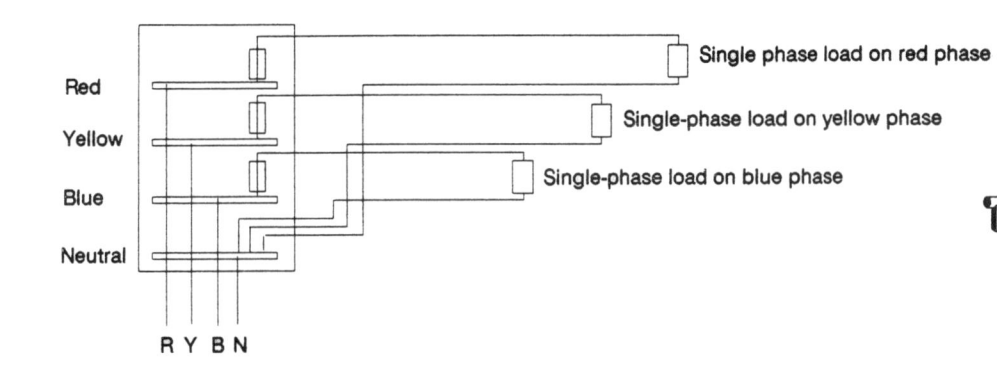

Single phase load on red phase

Single-phase load on yellow phase

Single-phase load on blue phase

Red

Yellow

Blue

Neutral

R Y B N

Diagram 2.20

In Diagram 2.20 there are three single-phase loads each connected to a different phase. If each of the loads is the same then the current in the neutral supplying the distribution board would be zero. As soon as any one load is switched off though, the loads in the three-phases are no longer balanced. This means there will now be a current flowing in the neutral conductor supplying the distribution board.

The current in the neutral conductor can be determined by drawing scaled phasor diagrams. First, consideration must be given to the relationship between each phase from the supply. As the angle between each phase is the same, there must be 120° between phases. Diagram 2.21

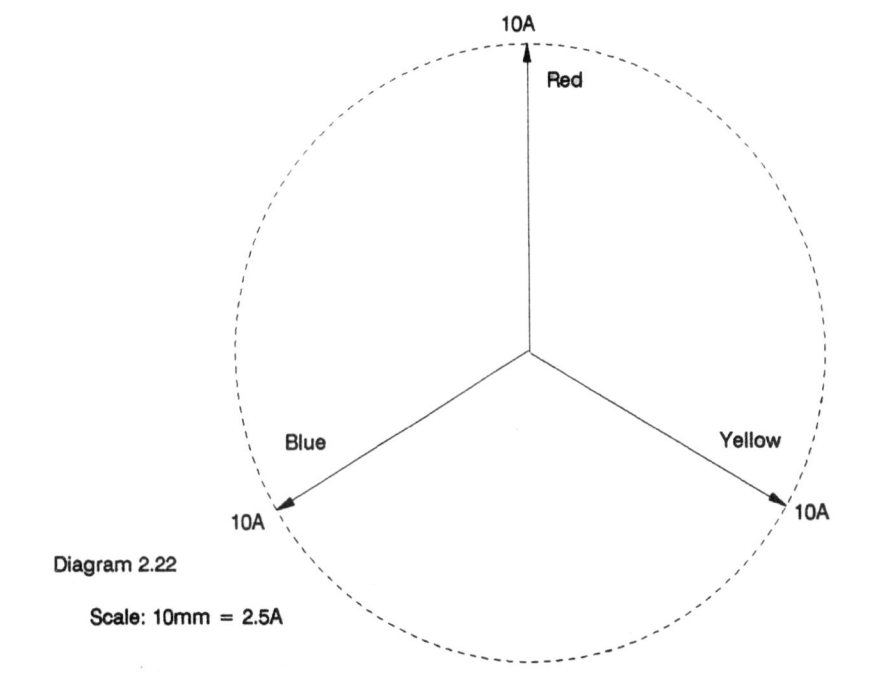

Diagram 2.21

Red phase

120°   120°

120°

Blue phase

Yellow phase

Assuming that the loads from the distribution board are purely resistive, such as heating elements, and no power factors need to be taken into account, a scaled phasor diagram can be constructed. The scale for Diagram 2.22 is 10mm = 2.5A and each phase has a load of 10A. Each line is at 120° to the other and is the same length. This can be checked by drawing a circle from the centre of the phasor diagram.

10A

Red

Blue

Yellow

10A

10A

Diagram 2.22

Scale: 10mm = 2.5A

By adding the phase loads together the current in the neutral conductor can be determined. Remember each line must be used at the original angle and at the original length. They must also be added in the direction of the arrows.

Add the red phase to the yellow.

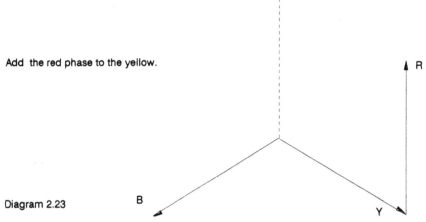

Diagram 2.23

In the same way the blue phase can now be added to the red.

If this is drawn accurately then the blue phase should end up pointing at the centre of the star.

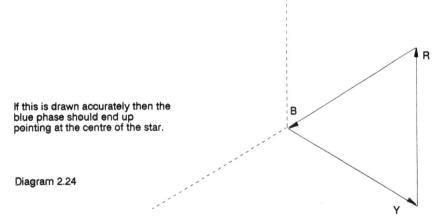

Diagram 2.24

As there is no distance between the centre point and the finish of the phasor addition, this indicates that there is no current flowing in the neutral conductor.

In the situation where the load in the red phase has been switched off but the yellow and blue still have 10A each, there will be a current in the neutral condutor. This can be determined by adding the yellow and blue phase loads and measuring the resultant.

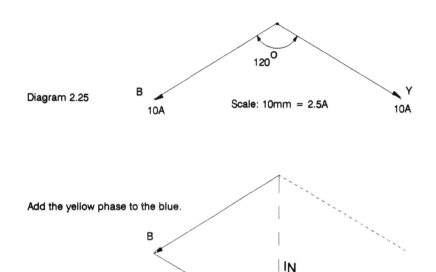

Diagram 2.25

Add the yellow phase to the blue.

Diagram 2.26

Now when the yellow phase is added to the blue there is a long distance between the arrow on the yellow to the centre or neutral point. A line drawn between these points will indicate the amount of current flowing in the neutral. In this case the line is 40mm so the current would be 10A.

That is basically how phasor diagrams can be used to determine the current flowing in a neutral conductor when there are three single phase loads connected to a three-phase 4 wire supply.

Question 5

Three resistive single-phase loads are connected to a three-phase distribution board. The loads are:

Red phase 12A
Blue Phase 5A
Yellow phase 7A

Using a scale of 10mm = 1A construct a phasor diagram and measure the current flowing in the neutral of the supply to the distribution board.

IR 12A

IY 7A

IN

IN = 6A

IB 5A   Scale 10mm = 1 ampere

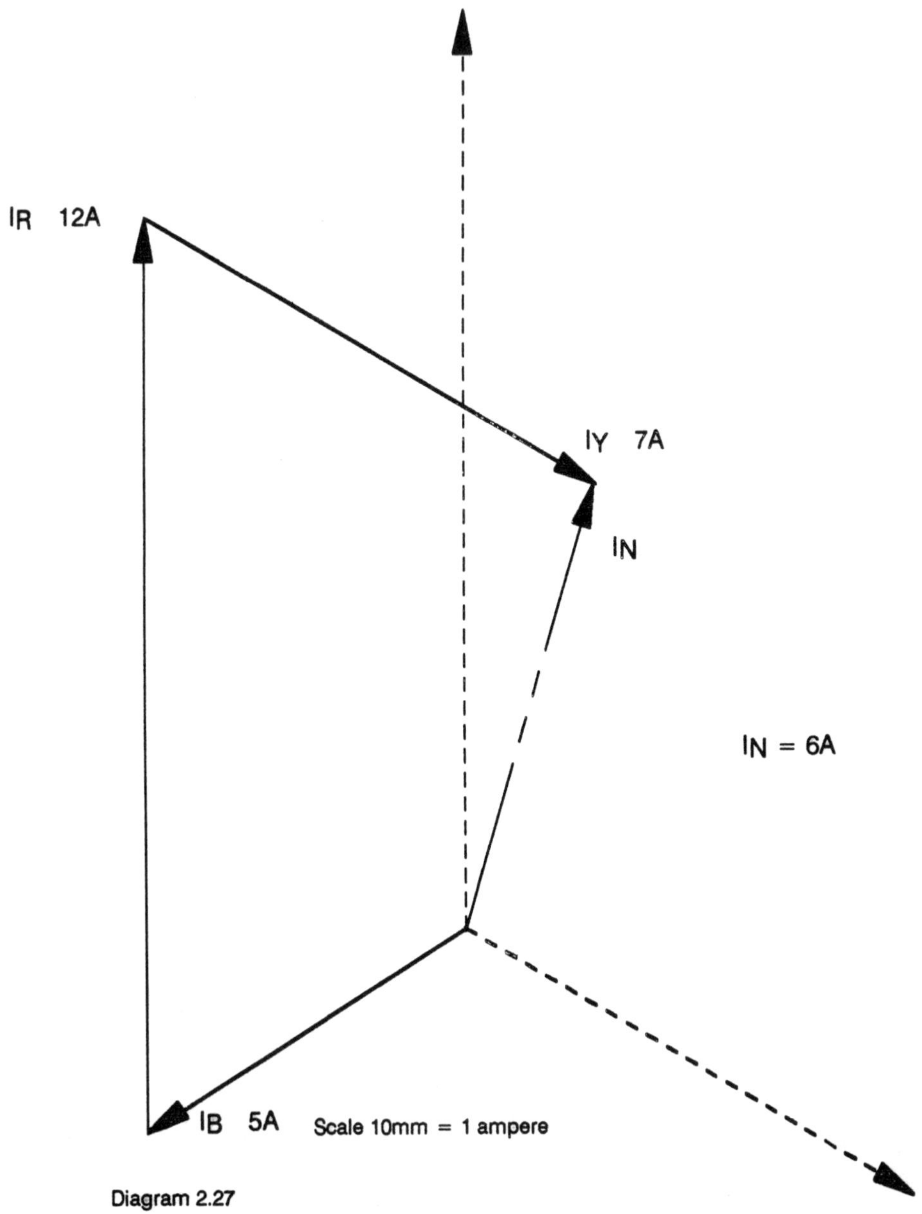

Diagram 2.27

Now try this - Five

A three-phase star connected transformer is loaded so that there is 60A drawn from the red phase with 40A from the yellow and 100A from the blue. Using a scaled phasor diagram determine the current that would be flowing in the neutral return to the star point of the transformer.

Power Supply Units

Supply units that are designed to convert a mains supply at 240V a.c. to a lower voltage d.c. have to include two basic stages. First the voltage has to be transformed down to the required value; then the alternating current must be converted into direct current. Diagram 2.28

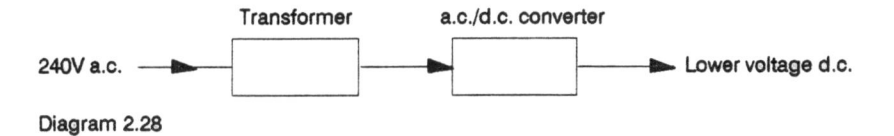

Diagram 2.28

The transformer would be a double wound type so that there is no electrical connection between the input and the lower output voltages. An autotransformer should NOT be used as it would be possible under some fault conditions to get mains voltage on the output terminals.

The a.c./d.c. converter is made up of a number of diodes. Each diode will allow current to flow easily when it is applied in the forward direction. But if the current is reversed the diode will oppose it. So if one diode is used on an alternating supply it will only allow half of the wave to flow. Diagram 2.29

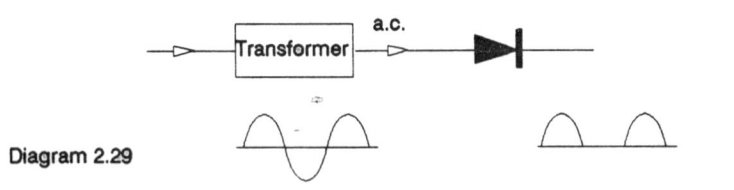

Diagram 2.29

This is now a direct current supply but because half of the wave has been lost its effectiveness has been reduced. Diagrams 2.30 and 2.31

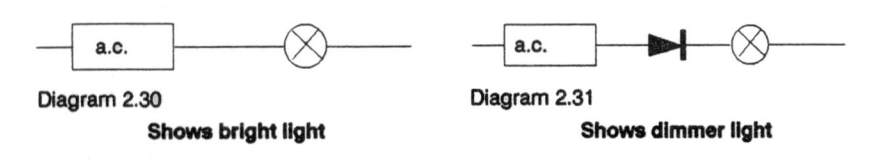

Diagram 2.30

Shows bright light

Diagram 2.31

Shows dimmer light

To get a d.c. supply and have full power a change-over arrangement is used. This is known as a rectifier bridge. Diagram 2.32

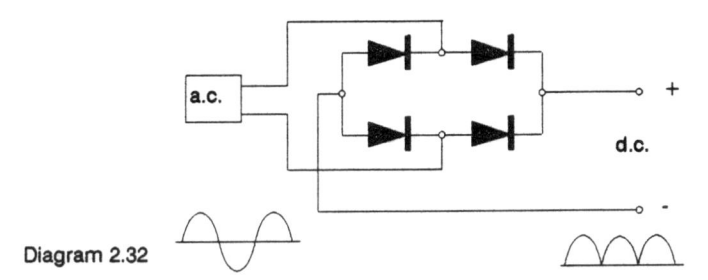

Diagram 2.32

The bridge has four diodes connected so they all point in one direction as shown in Diagram 2.32. They all point to the positive of the d.c. supply and away from the negative. The a.c. is connected between the other diodes. As the a.c. is flowing each diode will only allow current to flow in one direction and so all reversed current is blocked and only forward current is allowed through. This has the effect of reversing half of the a.c. waveform so that now it is all on one side of the zero line.

Measuring Current in a D.C. Circuit

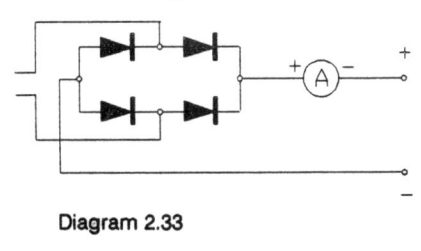

To measure the current flowing in the output an ammeter can be connected into one of the supply cables. Care must always be taken with d.c. instruments to ensure the polarity is correct. Diagram 2.33

Diagram 2.33

Measuring Voltage across a D.C. Circuit

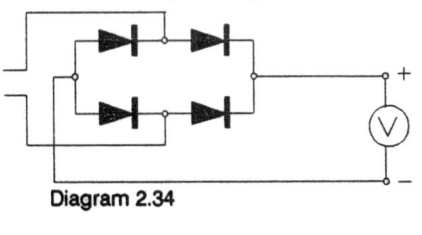

A d.c. voltmeter should be connected across the output if the voltage is to be measured, Diagram 2.34. Again it is important to connect the meter with the correct polarity.

Diagram 2.34

Question 6

The components shown in Diagram 2.1 are those required to produce a simple battery charger. Redraw these to create a circuit suitable for a battery charger unit. Show all d.c. polarities and label each component.

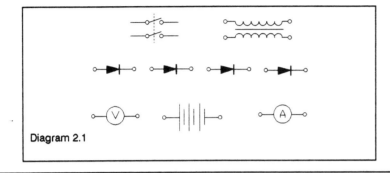

Diagram 2.1

Answer

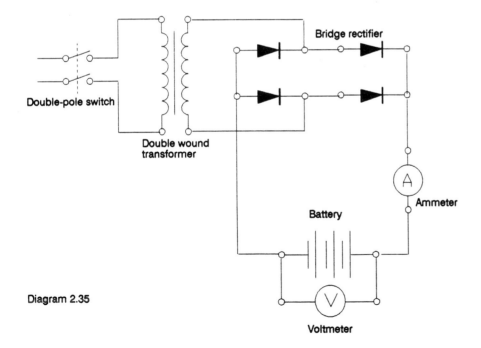

Bridge rectifier

Double-pole switch

Double wound transformer

Ammeter

Battery

Voltmeter

Diagram 2.35

Now try this - Six

Draw the output waveform of each of the folowing circuits.

a.

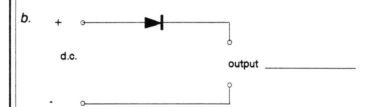

a.c.

output _____

b.

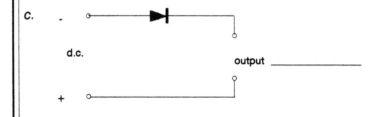

d.c.

output _____

c.

- ▸|▪

d.c.

output _____

+

Tips to help you answer the "Now try this" questions in this section.

Now try this - One

a) Notice that all four loads are single-phase so each will have two wires supplying it.

b) Remember that $P = UI \cos \emptyset$ where $\cos \emptyset$ is the power factor.

Now try this - Two

A diagram showing a TN-S system can be found in Part 2 of the IEE Wiring Regulations but don't just copy this - draw it in your own way.

Now try this - Three

Read the notes on Intake Distribution carefully and look at Diagram 2.12. If more information is still required this can be found in Section Four of Associated Electrical Installation Science published by CT Projects.

Now try this - Four

The answer to this should include references to mechanical damage, rewiring, type of flexible cables and limitations. Table 3A in Appendix 3 of the IEE On Site Guide may be of help.

Now try this - Five

Remember to keep the lines to the correct angles and use a scale that is large and simple to read.

Now try this - Six

Remember that the diode will only conduct when the polarity is correct. When the current is reversed no current will flow.

Further Revision

1. A 5 kW load takes a current of 40A when connected to a 240V supply. Calculate the power factor of the load.

2. A 415V three-phase supply is connected to a delta connected heating load. The rating of each of the three elements is 8kW. Calculate the current flowing in each supply cable.

3. Explain with the aid of a diagram, the earthing arrangements of a TN-C-S system.

4. A three-phase motor is supplied from a 6 way three-phase distribution board. If the distribution board is 10m from the main intake position draw a line diagram to show where it is necessary to install isolators.

5. Explain the advantage of a rising main busbar system over a conduit system to supply a number of suites of offices.

6. Draw a scaled phasor diagram and determine the current flowing in the neutral conductor when three single-phase loads are connected to a 415V 4 wire supply. The three loads are:

Red phase	- 4kW
Yellow phase	- 3kW
Blue phase	- 5kW

7. Draw the circuit diagram so that a 9V full wave rectified supply can be obtained from a 240V 50Hz supply using:

    1 double wound transformer with a centre tap on the
    secondary winding
    2 single diodes

Section Three

Wiring Systems

Questions covering aspects of the selection of cables and their uses can be found in this section.

1. Give examples of the wiring systems that would be suitable for the following installations.

   a. A domestic installation for lighting and socket outlets.
   b. An outside lighting installation which comprises of a number of tungsten halogen luminaires mounted on an outside wall. All cable runs must be on the outside surface.
   c. A fire alarm system.
   d. A motor driving a compressor.

   Give reasons for your answers.

2. Three electric heaters are to be installed in a sales area of a shop. Each heater is rated at 3kW, 240V and is to be wired with a separate pvc insulated and sheathed cable. The three cables are run together for a large part of their 15m length and go through an area with an ambient temperature of 30 °C. To ensure the maximum voltage drop is not exceeded no circuit should be greater than 5.5V.

   a. Determine the design current for each circuit and the rating of each protection device if BS 88 fuses are to be used.
   b. Calculate the minimum current rating of each cable.
   c. Select a minimum size of cable to comply with the current carrying capacity and voltage drop constraints.
   d. Determine the actual voltage drop when the cable is carrying the full 3kW load.

3. Determine the minimum cross sectional area of protective conductor from

$$S = \frac{\sqrt{(I^2 t)}}{k}$$

   for a 240V single-phase circuit which has the following:

   i) a value for Z_E of 0.3Ω.
   ii) a value for $R_1 + R_2$ of 0.7Ω
   iii) a circuit protective device of 30A to BS 1361.
   iv) a k factor of 143.

4. A fire detection system is to be installed in an office complex. Name a suitable fire detector device for each of the following situations and give reasons for your choice.

   a. general office areas
   b. staff canteen
   c. adjacent to fire exits
   d. computer rooms

Introduction

The wiring system is the heart of any installation. If the wiring breaks down the loads fail to work, and if the insulation fails dangerous conditions can arise. Selecting the correct system for the load, environment and cost is very important. Having carried out a basic selection it is then necessary to confirm the suitability by calculation.

Cable Selection

The selection of a particular type of cable for an installation depends on  many factors. These can include environmental conditions such as heat, damp, corrosive atmospheres and mechanical damage; or specification requirements that insist on low quantities of smoke, toxic and corrosive gases should a cable catch fire.

The actual selection is often a compromise between what is suitable and what is financially acceptable. For example mineral insulated metal sheathed cable (mims) is very suitable for wiring domestic premises but the cost would be many times greater than if pvc insulated and sheathed cable is used. There are occasions where cost must be given a low priority and safety given the maximum consideration. When supplying the pumps of a petrol filling station, for example, it would be extremely dangerous to use pvc insulated and sheathed cable. In fact it would also be against all safety recommendations which only allow for pvc sheathed mims and pvc/swa/pvc cables. Pvc in these cases can be substituted with other compounds such as XLPE (cross-linked polyethylene).

Regulations also dictate the type of wiring system in some special  installations. In Part 6 of the IEE Wiring Regulations under Agricultural and Horticultural Premises where it is expected that livestock may be present, the electrical equipment should be of Class II construction, or constructed of or protected by suitable insulating materials. Despite the fact that the environment is damp and open to mechanical damage, this rules out the use of wiring systems such as single cables installed in galvanised steel conduit.

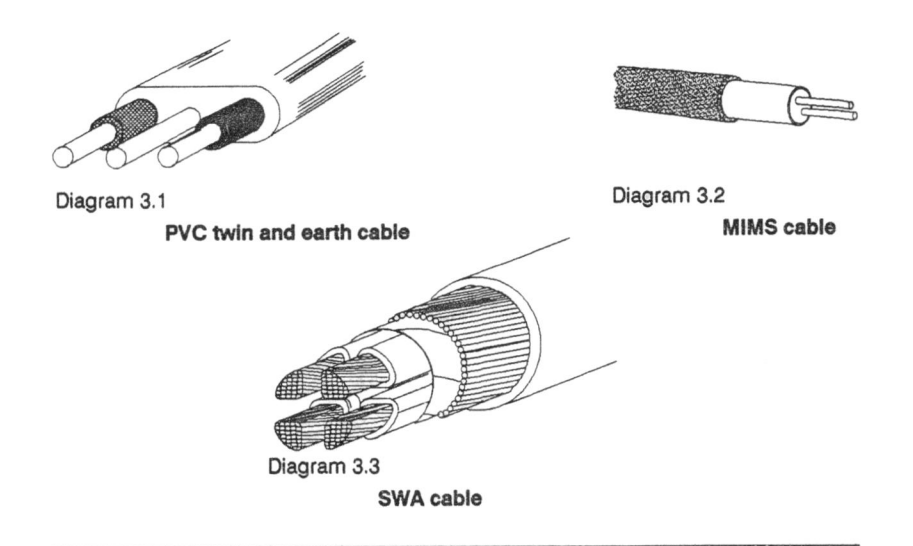

Diagram 3.1
PVC twin and earth cable

Diagram 3.2
MIMS cable

Diagram 3.3
SWA cable

Question 1

Give examples of the wiring systems that would be suitable for the following installations.

- a. *A domestic installation for lighting and socket outlets.*
- b. *An outside lighting installation which comprises of a number of tungsten halogen luminaires mounted on an outside wall. All cable runs must be on the outside surface.*
- c. *A fire alarm system.*
- d. *A motor driving a compressor.*

Give reasons for your answers.

Answer

a. Domestic installations would normally be wired in pvc insulated and sheathed cable. This would need to be protected from mechanical damage in some situations and run in positions where it is not exposed to ambient temperatures exceeding 30°C.

The reasons for this choice are that:

- it is comparatively cheap compared with other systems
- no special tools are required to terminate it
- installation can be quite fast as the cable is flexible and
- it can be concealed from view with the minimum of work.

b. There are several wiring systems that should be considered for the installation of outside lights. Whichever system is chosen it must be capable of withstanding all of the different weather conditions. This may exclude the use of pvc insulated and sheathed cable depending on the exact location.

If the wiring system is to be exposed to any mechanical damage then a galvanised conduit system with single cables installed in it may be best suited. Pvc insulated swa cable may also be suitable but where it is to be looped in and out of a number of tungsten halogen luminaires this may not look as neat as the galvanised steel conduit.

c. The choice of wiring system for a fire alarm system may depend on the installation requirements but in general, mims or FP200 cable would be used. It is important that fire alarm systems carry on operating as long as possible in the event of a fire. Both mims and FP200 cables are capable of operating even when they are exposed to direct flames.

d. Where a motor is driving a compressor there is going to be vibration. This means the wiring system chosen has got to be capable of withstanding this vibration without deterioration. Although mims cable can be used for this, flexible metal conduit with single-core pvc insulated cables is often more practical. It must be remembered that where flexible conduit is used, whether metal or not, a separate circuit protective conductor must be installed.

Now try this - One

What factors should be considered when selecting a wiring system for a paint store?

Load Calculations and Conductor Selection

It is important to ensure that any cable installed is capable of safely carrying the maximum load current in all conditions. Sometimes a total load is made up of several small loads. In this situation it must be determined what chances there are of all the loads being on all at the same time. Take, for example, the lighting in a domestic installation. It is very seldom that every light is on together. This is known as giving consideration to diversity and Table 1B in the IEE On Site Guide shows that you need only allow for 66% of the total current demand for the lighting load in a domestic installation. If, of course, you know this figure will be exceeded then the higher value must be allowed for.

Often the loading of equipment is given as the power in watts and this has to be converted into current in amperes before any cable capacity tables can be used.

Remember:

$$P = UI$$

so $\quad I = \dfrac{P}{U}$

There are other factors that must be taken into account before the tables can be used. It pays to follow a set sequence when doing this so that nothing is missed out.

1. Calculate the load current - where necessary take any power factor and efficiency into account.

2. Select an appropriate protection device - it must not be rated less than the load current.

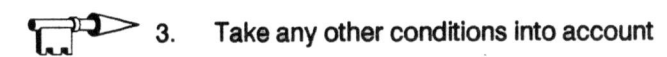 3. Take any other conditions into account

   - ambient temperature
   - grouping
   - thermal insulation
   - type of protection device

Let's look at an example

A single-phase motor is rated at 3kW at a power factor of 0.8 when supplied with 240V. The motor is protected by a fuse to BS 88. The pvc insulated cables are installed in conduit with one other circuit and the highest ambient temperature is $35^{o}C$.

1. Calculate the load current

   $$P = UI \ Cos \ \emptyset$$

   As we need "I"

   $$I \quad = \quad \dfrac{P}{U \ Cos \ \emptyset}$$

   $$= \quad \dfrac{3000}{240 \times 0.8}$$

   $$= \quad 15.6A$$

2. Select an appropriate protection device.

   The nearest fuse without being less than 15.6A for a BS 88 is 16A.

3.   Take any other considerations into account.

Ambient temperature

To find the correct factors for this Table 4C1 is used.
As we are using general purpose pvc cable the correction factor

for 35°C is **0.94**.

Grouping

Before Table 4B1 can be used the reference method for a conduit installation must be established from Table 4A. We assume that the conduit is on the surface so Method 3 applies.
The number of circuits we have is ours plus one other so from Table 4B1,

Reference Method 3, 2 circuits - a correction factor of **0.8**.

Thermal Insulation

As our installation does not encounter thermal insulation we use the factor of 1. If we needed to take this into account Table 52A and Regulation 523-04-01 would apply.

No thermal insulation - Factor **1**

Protection Device

This only applies to the use of BS 3036 devices where a factor of 0.725 must be used. As we are using

a device to BS 88 we can use the factor of **1**.

Now we can calculate the current to use when looking at the cable current rating tables.

For this calculation we must use the current that is the rating of the protection device as it is this current the cable will carry until the device operates.

$$\frac{\text{current rating of device}}{\text{factor for temp. x factor for grouping x factor for thermal insulation x factor for protection device}}$$

$$\text{or} \quad \frac{I_N}{C_a \times C_g \times C_i \times C_f}$$

$$= \frac{16}{0.94 \times 0.8 \times 1 \times 1}$$

$$= \quad 21.28A$$

The next stage is to find a cable capable of carrying the current of 21.28A. This of course means the cable rating in the tables must not be less than 21.28A.

There are again a number of stages to go through.

  i) Select the correct table
- core arrangement and
- type of insulation

  ii) Select the appropriate columns
- Reference method
- number of cores
- single or three-phase

Using the Tables in Appendix 4 of the Wiring Regulations or Appendix 7 of the On Site Guide the appropriate Table for our example can now be selected. This is Table 4D1A or Table 7A1 for single core pvc cables. The columns that apply are those under Reference Method 3 and in particular Column 4 which applies to 2 cables single-phase. In this column the first cable rating above 21.28A is 24A and from column 1 we can see that this is for a 2.5mm^2 conductor.

There is one more stage that must be checked before we can be sure this conductor size is suitable. This is to make sure that when the motor  is working on full load the voltage drop is not such that the motor would not be able to work correctly. Before we can determine the voltage drop there are two more details we need to know about our motor circuit. These are, the length of cable run and the minimum voltage that the motor will work on. The length of run we will assume is 10 metres and the minimum voltage is 235V, giving, a maximum voltage drop of 5V on full load.

Details of the voltage drop for a 2.5mm^2 cable for Reference Method 3 can be found on Table 4D1B or 7A2 col 3. This is 18mV for a 2.5mm^2 cable for every ampere that flows through every metre of cable. We have 21.28A and 10m, so the actual voltage drop is

$$18 \times 21.28 \times 10 = 3800mV$$

$$\text{or} \quad \frac{3800}{1000} = 3.8V$$

As this is under the 5V maximum we find the 2.5mm^2 conductor under these conditions is acceptable.

Although this does appear to be a lot of work in looking different things up it is important to complete all stages in the correct order. It can be dangerous not to check all the factors before a cable is installed.

Check the following:

Factors to be considered

Environmental
  temperature
  corrosive
  moisture
  mechanical

Cable Run
  length
  other cables
  thermal insulation
  method of installation

Cost
  environment
  safety requirements

Cable considerations

Temperature
  pvc general purpose
  mims
  pvc/swa/pvc

Mechanical
  pvc sheathed
  metal sheathed mims
  FP200
  steel wire armoured
  steel conduit pvc conduit

Moisture
  terminations

Question 2

Three electric heaters are to be installed in a sales area of a shop. Each heater is rated at 3kW, 240V and is to be wired with a separate pvc insulated and sheathed cable. The three cables are run together for a large part of their 15m length and go through an area with an ambient temperature of 30°C. To ensure the maximum voltage drop is not exceeded no circuit should be greater than 5.5V.

  a. Determine the design current for each circuit and the rating of each protection device if BS 88 fuses are to be used.
  b. Calculate the minimum current rating of each cable.
  c. Select a minimum size of cable to comply with the current carrying capacity and voltage drop constraints.
  d. Determine the actual voltage drop when the cable is carrying the full 3kW load.

Answer

a. Design current of each circuit

$$= \frac{3000}{240} = 12.5A$$

The protection device to BS 88 for 12.5A = 16A

b. The minimum current rating of each cable

$$\frac{\text{protection device rating}}{\text{correction factors}}$$

Relevant correction factors:

    grouping - 3 circuits touching   = 0.79
    temperature = 30°C   = 1

$$I_z = \frac{I_N}{C_a \times C_g} = \frac{16}{0.79 \times 1} = 20.25A$$

c. Minimum cross sectional area of conductor

Table 41D4A col 2		Table 4D4B col 3
1.5mm^2	21A	29mV

A 1.5mm^2 conductor is large enough for the current rating.

Checking for voltage drop
    length of run   = 15m
    current   = 12.5A

$$\text{voltage drop} = \frac{29 \times 15 \times 12.5}{1000} = 5.43V$$

d. The actual voltage drop is 5.43V.

Now try this - Two

A 415V three-phase 9kW water heater is to be wired with pvc insulated single conductors installed in steel conduit. The water heater is 15m from the distribution board at the main intake position. Overcurrent protection is be HBC fuses to BS 88. For 5m of the run a second similar circuit is installed in the same conduit. Calculate the minimum cross sectional area of cable suitable if the maximum voltage drop must not exceed 5V.

The Earth Fault Path

The earth fault path consists of the circuit protective conductor and the earth resistance external to the installation.

To complete the circuit the phase conductor is also included in the resistance path.

As the circuit external to the installation is in the hands of the supply company this is all put together as one value covering the impedance of the earth circuit, transformer winding and phase conductor. This impedance is represented by Z_E, the external impedance.

The cross sectional area of the phase conductor is determined, as we have seen, by the current demand of the load. The resistance of this is referred to as R_1.

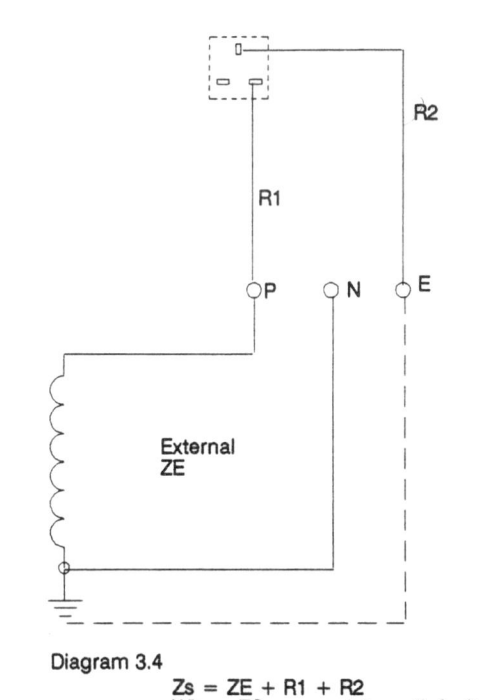

External ZE

Diagram 3.4

$Z_s = Z_E + R_1 + R_2$
Where ZS = complete earth fault path

The cross sectional area of the circuit protective conductor cannot be determined in the same way as the live conductors, for it only comes into use under fault conditions. Then it has to help clear the fault current as fast as possible.

The maximum time that a fault can be cleared in depends on the type of circuit and where it is installed. In some circumstances it is safe to have faults flowing for up to 5 seconds. In other situations it is essential to have them cleared in 0.2 seconds. As the resistance of the phase conductor has already been determined and the external impedance is outside of our control, it is often the resistance of the circuit protective conductor that is used to determine the value of prospective earth fault current.

To keep the selection of cpc's simple Table 54G in the Regulations gives a quick guide. Assuming the circuit protective conductor is copper, the same as the phase conductor, then for circuits with a phase conductor no greater than $16mm^2$ the cpc would be the same size. When the phase conductor is between $16mm^2$ and $35mm^2$ the cpc need be no greater than $16mm^2$. For circuits where the phase conductor is greater the $35mm^2$ the cpc need be no larger than half that of the phase. In practice this will need rounding up to the next actual size available. Like most quick and easy guides the answer obtained is not always the most cost effective. The same applies with this. To get a more accurate calculation Regulation 543-01-03 should be consulted and the formula given there, applied.

$$S = \frac{\sqrt{(I^2 t)}}{k}$$

where

S = the cross sectional area of the conductor

I = the current required to flow under fault conditions

t = the time the protection device will take to operate when the fault current is flowing

k = the relevant factor found in Tables 54B to 54F.

This formula takes into account the fact that heat is produced when a fault current flows and this heat can damage the insulation of the cable if it is not cleared within the appropriate time.

Question 3

Determine the minimum cross sectional area of protective conductor from

$$S = \frac{\sqrt{(I^2 t)}}{k}$$

for a 240V single-phase circuit which has the following:

i) a value for Z_E of 0.3Ω.
ii) a value for $R_1 + R_2$ of 0.7Ω
iii) a circuit protective device of 30A to BS 1361.
iv) a k factor of 143.

Answer

To calculate the minimum cross sectional area of protective conductor the following formula must be used.

$$S = \frac{\sqrt{(I^2 t)}}{k}$$

"I" can be calculated from

$$I_a = \frac{U_o}{Z_s}$$

where I_a = current under fault conditions
U_o = supply voltage
Z_s = complete earth fault path

Z_s can be calculated from

$$Z_s = Z_E + R_1 + R_2$$
Z_E = 0.3Ω
$R_1 + R_2$ = 0.7Ω

so Z_s = 0.3 + 0.7
= 1.0Ω

$$I_a = \frac{U_o}{Z_s} = \frac{240}{1.0} = 240A$$

"t" can be determined from the time/current characteristics for a 30A BS1361 fuse shown in Fig.1 Appendix 3 of the Regulations.

t = 30A fuse 240A fault current
= 0.2 seconds

using $$S = \frac{\sqrt{(I^2 t)}}{k} = \frac{\sqrt{(240^2 \times 0.2)}}{143} = \frac{\sqrt{11520}}{143}$$

$$= \frac{107.33}{143}$$

$$= 0.75mm^2$$

This has to be rounded up to the nearest size of cable, which is 1.0mm^2.

Now try this - Three

A 240V circuit is wired in a 6mm^2 pvc/swa/pvc cable and is protected by a 50A BS 88 general purpose fuse.

a. What is the maximum value of Z_s this circuit can have if the disconnection time must not exceed 0.4 seconds?
b. Determine the maximum earth fault current.

Fire Detection Devices

It is important to consider the devices that are used on fire detection circuits. These often require special circuits that give the maximum protection in the event of a fire. The detection devices generally fit into three categories, smoke, heat and manual. The first two are of course automatic and can detect fire for 24 hours a day throughout the year, whereas the third requires somebody being present and setting off the alarm. There are other automatic systems in use but these usually require specialist installation.

Smoke detectors, as the name implies, detect the presence of  smoke in the air. There are generally two types in use, the ionised chamber and the optic chamber. The use of both is very similar and it is the ionised chamber type that has become widely used in domestic installations. They should not be used where there is the possibility of smoke or dust in the air under normal working conditions as this can set them off accidentally.

Heat detectors also fall into two general types, those that use  bimetallic arrangements and rely on a movement taking place to make or break contacts, or the ones that operate with the action of a thyristor.

The bimetallic type have sometimes been prone to problems due to mechanical defects. Both can be affected by the sudden rise in temperature which is not necessarily anything to do with a fire. "Rate of rise" detectors can take into account changes in temperature within an area without setting off alarms.

 Manual units such as break glass switches can be effective as the human body can often detect fire as soon if not before automatic devices. They are however prone to abuse by being set off when there is no fire and being inaccessible when there is one.

The siting of all detection devices is important as this may determine whether they will be activated when the need arises. Diagrams 3.8 and 3.9.

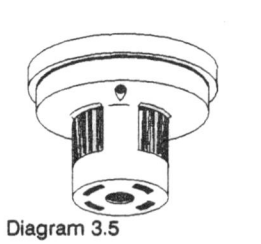

Diagram 3.5

Smoke detector

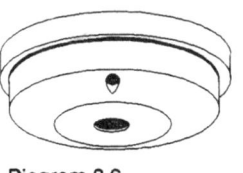

Diagram 3.6

Heat detector

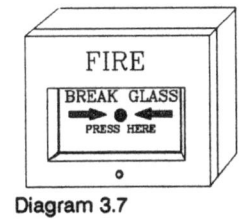

FIRE

BREAK GLASS

PRESS HERE

Diagram 3.7

Manual unit

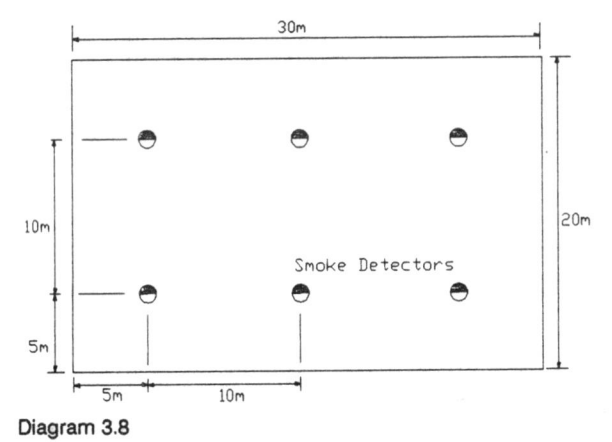

Diagram 3.8

Siting of smoke detectors on a flat ceiling

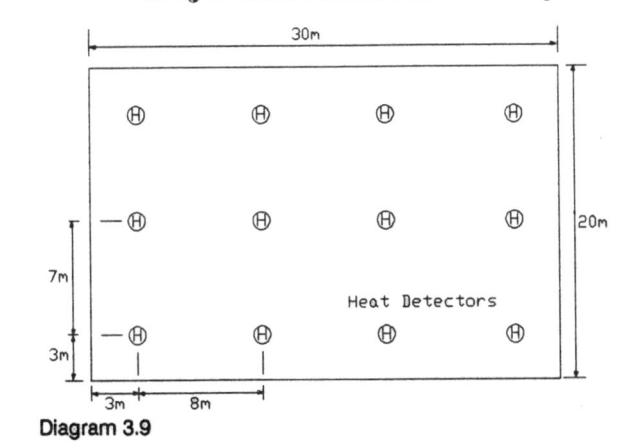

Diagram 3.9

Siting of heat detectors on a flat ceiling

Question 4

A fire detection system is to be installed in an office complex. Name a suitable fire detector device for each of the following situations and give reasons for your choice.

  a. *general office areas*
  b. *staff canteen*
  c. *adjacent to fire exits*
  d. *computer rooms*

Answer

a. The fire detection within a general office may depend on whether smoking is permitted in the area. If smoking is permitted and the area is likely to become very smoky in normal use then heat detectors would be best. If, however, the office is a smoke free zone then a smoke detector would probably be advisable. This is due to the fact that many offices use a quantity of electrical equipment and it can be this when a fault develops that causes fires. When things such as waste paper baskets catch fire then either detection device would be suitable. In addition to the automatic devices, manual break glass contacts should be sited on all exit routes.

b. The staff canteen can be a difficult problem for it is often very smoky and the temperature varies in there throughout the day. Generally a "rate of rise" type heat detector is most suitable to limit any false alarms.

c. All fire exits should have manual activated devices as these routes are going to be used if the building is occupied.

d. Computer rooms are generally dust free environments. The smoke detector is usually most suitable but there are often more sophisticated systems in use as well due to the nature and value of the equipment.

Now try this - Four

Explain, with the aid of a diagram, how detection devices are wired on a "closed circuit" system.

Tips to help you answer the "Now try this" questions in this section.

Now try this - One

Remember that a paint store may contain fumes that could become a hazard.

Now try this - Two

Remember to look up first:

- the reference method for the installation of cables Table 4A

- grouping factor Table 4B1

Calculate the line current - remember $P = \sqrt{3}\, UI$

Look up a suitable BS 88 protection device.

Calculate the maximum load current for the cable:

$$\frac{\text{rating of protection device}}{\text{correction factors}}$$

Look up a suitable cable Table 4D1A

Check the voltage drop Table 4D1B

Check against 5V.

Now try this - Three

Remember Z_S values can be found in the IEE Regulations in Tables 41B1, 41B2 and 41D; and Tables 2A, 2B and 2C in the On Site Guide. Regulation 413-02-08 can help with Ia.

Now try this - Four

Remember that a relay is required in this type of circuit and that BS 3939 drawing symbols should be used.

Further Revision

1. Select a suitable wiring system for the following areas associated with petrol filling stations. Give reasons for your choice.

   a. sales shop which is outside any hazardous area
   b. canopy lighting 3m above the top of the pumps
   c. pump supplies

2. A two core pvc/swa/pvc cable is to be installed underground to supply a new office remote from the main building. The cable is to be protected by an m.c.b. to BS 3871 Type 2 in a distribution board 30m from the new office consumer unit. The loads in the new office will be:

Lighting	10 twin 60W fluorescent luminaires
	5 filament luminaires
Power	2 ring final circuits protected by 32A fuses to BS 88
Heating	2 3kW thermostatically controlled fan heaters
	1 4.5kW storage radiator
	1 3kW thermostatically controlled water heater

   Calculate the minimum cross sectional area of cable suitable for this installation assuming that the voltage drop must not exceed 3V.

3. A radial circuit supplying a 13A socket outlet is wired in 2.5mm^2 cable which has a 1.5mm^2 circuit protective conductor incorporated within the sheath. Calculate

   i) the R1 and R2 values if the cable is 18m long
   ii) the actual Zs value if Ze is 0.3Ω
   iii) the maximum Zs value if a 20A protection device to BS 88 parts 2 & 6 is used.

4.a. What is the minimum fault current that needs to flow through a 100A fuse to BS 88 parts 2 & 6 if it is to operate in 0.4 seconds?

   b. If the protection device in (a) was replaced with a fuse to BS 3036 when the same fault current was flowing how long would it take the new fuse to operate?

5. Draw a block diagram showing how the control for a space heating system would work. The controller is supplied with information from a room thermostat, cylinder thermostat and programmer. It has to control the fuel supply to the boiler and the pump to the hot water.

Section Four

A.C. Motors

Questions to be answered in this section.

1.a. Explain how the rotating magnetic field is produced in a single-phase split-phase motor.

b. Using circuit diagrams explain how the direction of rotation can be reversed in a single-phase split-phase motor.

2. Identify the numbered parts of the three-phase motor shown in Diagram 4.1.

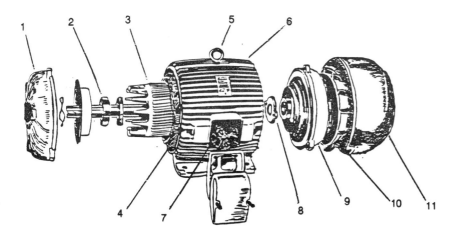

Diagram 4.1

3. Draw the circuit diagram for a 415V three-phase motor started from a direct-on-line starter with a 240V coil. Label all parts.

4. A motor has a voltmeter, ammeter and wattmeter connected to its supply cables. When running at full load the following readings were obtained:

volts - 238V
amperes - 32A
watts - 5kW

a. Determine the power factor.
b. Determine the current that would flow through a capacitor circuit if the power factor was corrected to unity.

5.a. A 415V three-phase motor is found to have an efficiency of 85% on full load. The power factor at this time is 0.8 lagging when the full load is 5.6kW. Calculate the current of the motor under these conditions.

b. If an eight pole induction motor runs at 12 rev/sec and is supplied from a 50Hz supply calculate the percentage slip.

Introduction

There are a number of different a.c. motors that should be given consideration when dealing with this subject.

These include:
- single-phase (split phase)
- single-phase capacitor start
- universal (series)
- three-phase cage induction
- three-phase wound rotor

There are several aspects to be aware of when looking at these. First it is important to have an idea of their basic construction. This does not mean every nut and bolt, but the main parts such as the type of rotor and stator. It is also necessary to know the circuit diagram of the internal connections of the windings. Again this need not be complex but simply what is connected to what, using standard symbols.

It is also important to look at the control of motors. This will include calculations relating to full load currents and motor speeds.

The Workings of A.C. Motors

To follow how an a.c. motor works consideration must be given to how alternating current is generated. This can be described as a magnet turning inside three coils.

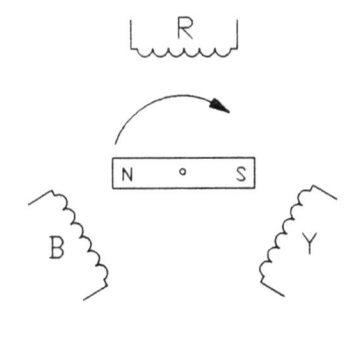

Diagram 4.2

As the magnet revolves its poles go past each coil in turn. As the magnetic field is moving, an e.m.f. is induced in each coil. The coils are effectively positioned 120° apart so the e.m.f.'s are induced at different times and three voltages at 120° intervals are produced.

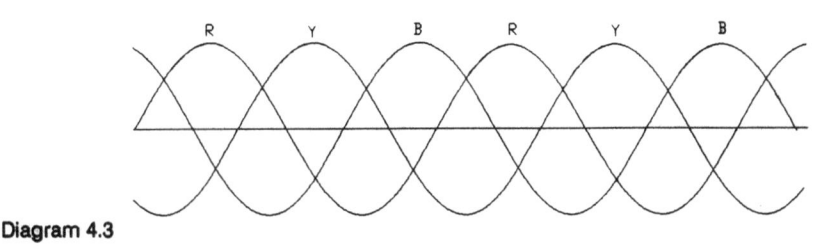

Diagram 4.3

The pattern produced always follows the sequence Red, Yellow, Blue.

Now, relating this to a three-phase motor that is being supplied with the generated output, it can be considered as a generator in reverse.

Diagram 4.4

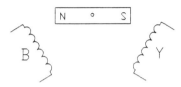

If the coils are connected up in the same sequence then the magnet will spin in the same direction. So in this case it is clockwise for the sequence RYB. If you take this out to a number of coils the sequence can become clearer.

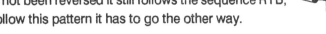

R Y B R Y B R Y B R Y B

If any of the coils are swapped over the sequence is reversed. Changing Red and Yellow.

Y R B Y R B Y R B Y R B

As the generator has not been reversed it still follows the sequence RYB, so for the motor to follow this pattern it has to go the other way.

Y R (B Y R) (B Y R) (B Y R) B

The direction of rotation has now been reversed.

This works well for three-phase motors but single-phase does not have a sequence to follow. Single-phase is the output from just one of the generator coils.

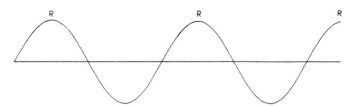

Diagram 4.5

 One a.c. sine wave does not create a moving magnetic field that is sufficient to start a motor. This means that a second phase has to be produced inside the machine.

There are many examples in a.c. equipment where highly inductive devices are used and a lagging phase angle is created. An inductive device consists of coils of copper wire wound around a laminated iron core. Two factors affect the total inductive effect of the device. These are the resistance of the coil and the amount of iron surrounding it. The less the resistance of the coil the greater the effect of the inductance. If the coil is deep into a laminated iron core the inductance is greater than if the coil is in air. In theory a coil with no resistance and an ideal iron core can create a phase shift of 90°. This means that the current through such a coil would occur 90° after the current through a coil with no inductance. This is, of course, also 90° behind the voltage waveform.

Now relate this to a single-phase motor. A thick conductor will have less resistance than a thin one. An example is that a $6mm^2$ cable has less resistance than a $1.0mm^2$ one. So by having two windings, one of thin wire and the other of thick, there is one with a low resistance the other with a higher resistance. As already seen a coil embedded in a laminated iron core has more inductance than one in air. So now the two coils have to be fitted into the stator. The thick coil with low resistance is to be the one that has the greater inductance so this is fitted deep into the slots on the stator. This is now almost totally surrounded by the laminated iron core. The coil with the thinner wire is fitted closer to the surface of the stator so that it has less iron surrounding it.

Now there is a coil with low resistance surrounded by a laminated iron  core. When current is passed through this the inductance makes the current lag by some angle less than 90°. The second coil has a high resistance and far less laminated iron core around it so when current is passed through the coil it only lags by a small amount behind the voltage. The overall effect is that the two coils have a phase displacement between them of about 30°.

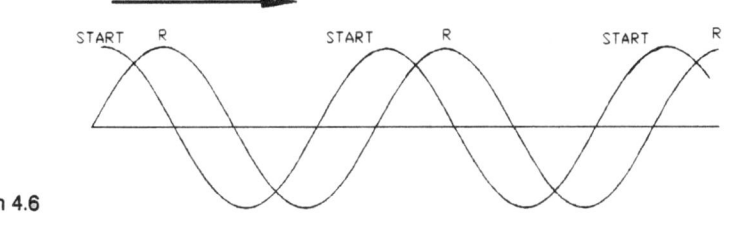

Diagram 4.6

This is enough to start a motor turning. Once the motor is up to speed the two windings are no longer required and the motor will continue to run on just one winding. As the thinner of the two windings will probably not carry the load current indefinitely this is switched off. The switch for this is mounted inside the motor and automatically opens the circuit to this winding at a predetermined shaft speed. As this winding is only used to get the motor up to speed it is known as the start winding.

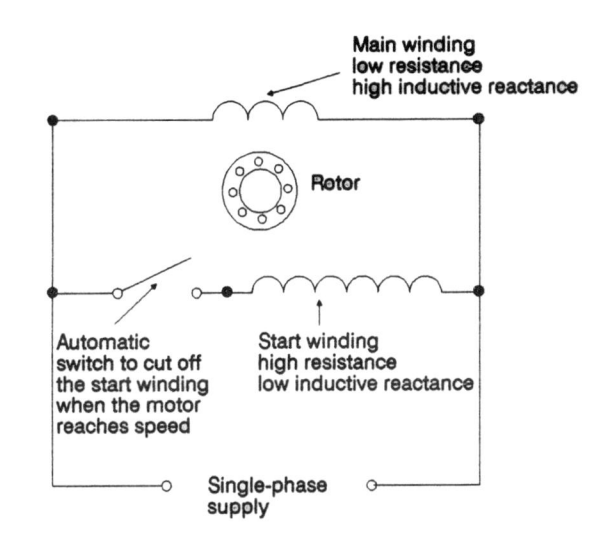

Diagram 4.7

The direction of rotation of a single-phase motor depends on the direction it is started in. Assuming that the direction in Diagram 4.7 is clockwise, then if the connections to the start winding are reversed the motor will rotate in an anti-clockwise direction when started.

Just changing the supply connections would have no effect on the direction of rotation because both coils have then been changed over and they are back where they started.

To create a larger "shift" between the two windings a capacitor is used in series with the start winding. This is then known as a capacitor start split-phase motor.

Question 1

a. *Explain how the rotating magnetic field is produced in a single-phase split-phase motor.*

b. *Using circuit diagrams explain how the direction of rotation can be reversed in a single-phase split-phase motor.*

Answer

a. The single-phase supply by itself will not produce a rotating magnetic field to get the motor started. A second winding has to be put in the stator core, connected in parallel with the first.

The main winding is made of thick wire and is placed deep in the iron core. The second coil is made of thinner wire and placed nearer to the surface of the iron core. As the main winding has less resistance and more reactance than the other coil a phase shift is produced. This in effect produces two-phases inside the motor.

When the motor is switched on the phase difference in the two windings is enough to produce a rotating magnetic field to get the motor started. As the motor reaches full speed a centrifugal switch disconnects the second (start) winding as this is no longer required.

b. To reverse the direction of rotation of a single-phase motor the connections to either the start or run winding can be changed over, but not both.

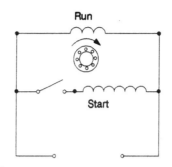

Diagram 4.8

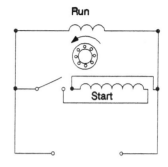

Diagram 4.9

Now try this - One

Explain using diagrams how the direction of rotation can be reversed on a three-phase cage rotor induction motor.

The Construction of a Three-Phase Induction Motor

A simplistic way to describe the action of a three-phase motor is to compare the rotor to a permanent magnet. This was described in the background to the previous question. This is in fact not true when it comes to the construction of the motor. There is no permanent magnet, in fact when the machine is switched off there is no magnetic field at all. It can get even more puzzling when you consider there is no electrical connection at all to the rotor. This means you can't put a supply on to it to make it an electromagnet. So how does it become magnetised?

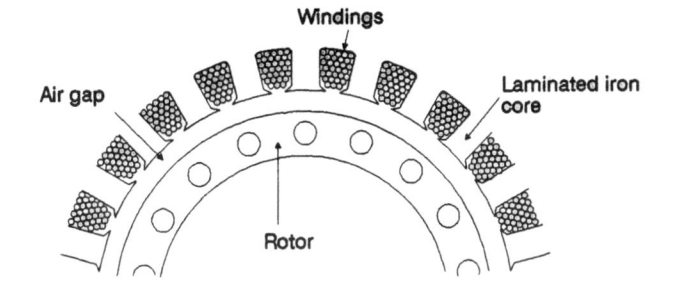

Diagram 4.11

Cross-section of an a.c. motor

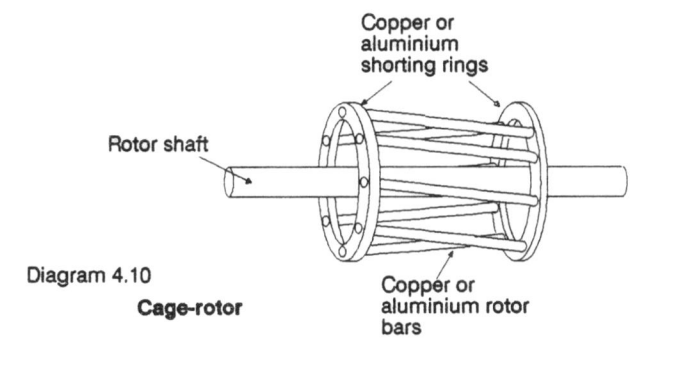

Diagram 4.10

Cage-rotor

The rotor of an induction motor is made up of a cage of copper or aluminium bars, as in Diagram 4.10.

In practice these are not so easy to see as they are embedded in the laminated iron core. If you think of each of these copper or aluminium bars as being an electrical circuit and they are all shorted out by the end rings then we can look and see how it works.

When the motor is assembled each rotor bar is close to a coil on the stator. Diagram 4.11

The stator coils are supplied with an alternating current which creates a moving magnetic field. This magnetic field not only goes around the coil but through the iron core across the air gap and into the rotor bars. This is the same principle as that for a double wound transformer. In this case the primary winding is the stator coil and the secondary is the rotor bars. As the rotor bars are all shorted out with the end rings the resistance of the circuit is low and the induced e.m.f. from the stator coils produces high currents in the rotor. This induced current in turn produces its own magnetic field. The magnetic field in the stator and the magnetic field in the rotor interact with each other and the rotor is moved. As the magnetic field in the stator moves with the supply voltage the rotor follows it and continues to rotate.

Question 2

Identify the numbered parts of the three-phase motor shown in Diagram 4.1.

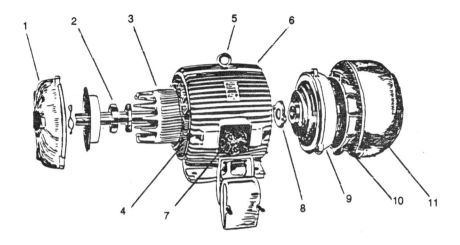

Answer

1. End shield, drive end
2. Ball bearing, drive end
3. Cage rotor
4. Stator windings
5. Eyebolt
6. Stator frame
7. Terminal box
8. Bearing cap
9. End shield, non drive end
10. Fan
11. Fan cover

Now try this - Two

Name the material each of the following will be made from in a three-phase cage rotor induction motor:

1.   Stator windings

_____

2.   Stator frame

_____

3.   Rotor shaft

_____

4.   Cooling fan

_____

5.   End shields

_____

6.   Rotor bars

_____

7.   Rotor core

_____

8.   Ball bearings

_____

9.   Terminal box

_____

10. Bearing caps

_____

Motor Starting

Most a.c. motors, if connected across the appropriate supply, will work. Unfortunately there are a number of reasons why in most cases it is not as simple as that. For a start the IEE Wiring Regulations tell us that all  motors over 0.37kW must have control equipment that protects against overload of the motor and automatic restarting after a stoppage or voltage failure. In addition to this the Supply Companies put restrictions on the size of motor that can just be switched directly on. This means that these three points must be taken into account when starting motors.

First let's look at two methods of overcurrent protection used in starters. Both are connected in series with the main supply cables to the motor so all current has to flow through them.

The first works on the principles of magnetism. It is basically a coil which becomes magnetised when a certain amount of current flows through it. Diagram 4.12. Inside the coil is a steel plunger which is pulled through the coil when the current reaches the critical value. When it is pulled right through it operates some contacts which switch off the motor. When motors start off from a stationary start they take a lot more current than when they are running. This current would be high enough to cause the overload trip to operate. To overcome this an oil dashpot is fitted to the bottom of the plunger so that when the starting current tries to pull it the oil slows down the rate of travel. If the high current continues, as when a fault develops, the plunger will eventually pull through the oil and operate the contacts.

There are two adjustments on this type of device, (i) there is a hole in the piston which can be made larger or smaller, thus allowing the oil to pass through at different rates, (ii) the bath of oil can be screwed up or down altering the length of travel for the plunger. Maintenance is required on this type of device to ensure the oil is kept at the correct level.

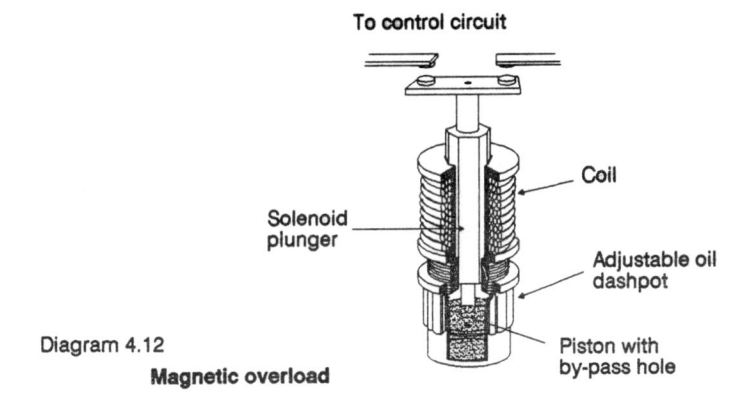

Diagram 4.12

Magnetic overload

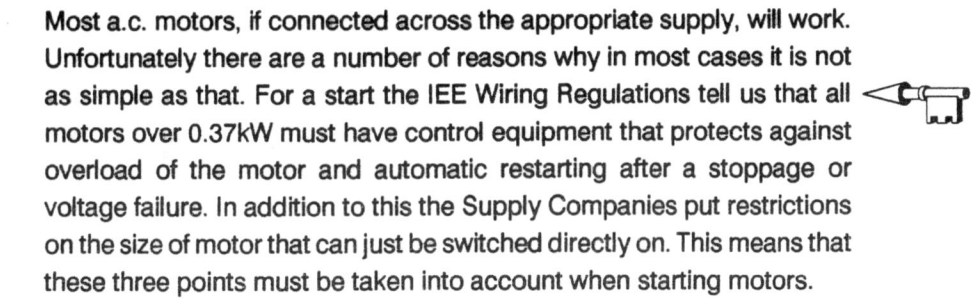

The second method of overload protection works on the heating effect of current flow. A wire is wrapped around a bimetallic strip. Diagram 4.13 When the motor takes excessive current the wire heats up causing the strip to bend and operate contacts. Starting currents are not the same problem with this method, as the heating effect takes longer and so the current has dropped to its running value before the trip can operate.

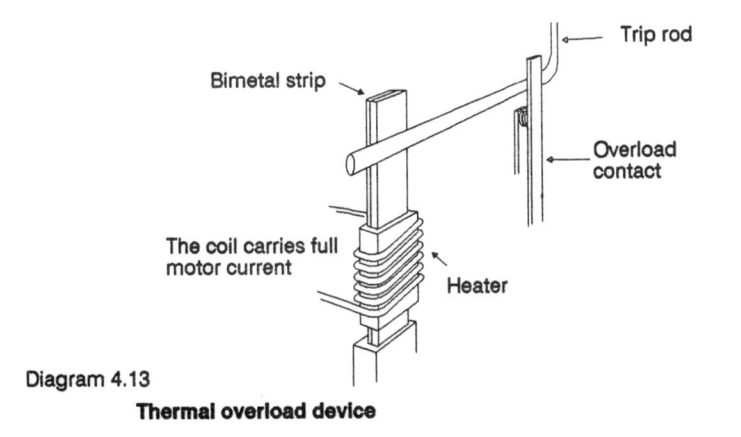

Diagram 4.13

Thermal overload device

The control circuit for most motor starters is similar. A start button 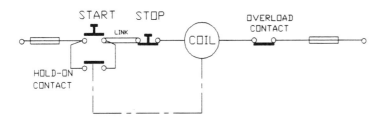 completes a circuit which energises a coil. Diagram 4.14. The coil pulls in the main contacts to supply the motor and also an auxiliary "hold-in" contact that shorts out the start button so the motor continues to work after the start button is released.

Diagram 4.14

Control circuit for direct-on-line starter

The stop button and overload contacts are In series with the coil . Should either of these be opened the motor will stop and not start again until the start button is pressed. This meets the requirements with regard to protection against automatic restarting after stoppage or voltage failure.

On occasions it is necessary to have extra stop and start buttons away from the starter. These can be connected to the control circuit as shown in Diagram 4.15.

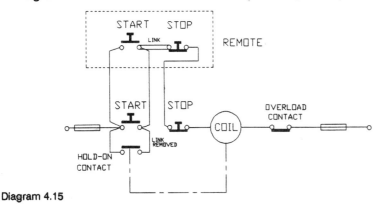

Diagram 4.15

Control circuit with remote start/stop

Where a motor can be started directly onto the supply a "direct-on-line" starter is used. This is basically a three- phase contactor with the control circuits incorporated into it.

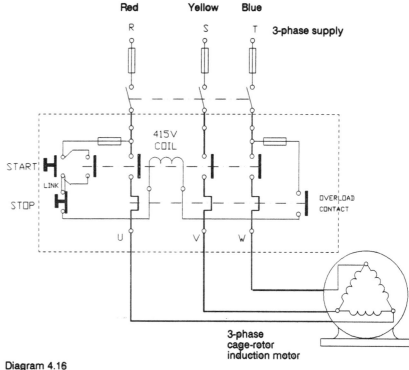

Diagram 4.16

Direct-on-line starter

Power (main) circuit shown by thick lines.

Control (auxiliary) circuit shown by thinner lines.

The fuses provide short-circuit protection.

The overloads may be thermally or magnetically operated.

The control circuit sometimes has to work from a 240V supply even though it controls a 415V three-phase motor. In this case a neutral line is required and the control circuit returns through this instead of one of the phases.

Where a motor cannot be started direct-on-line a method of current reduction must be used. The most common of these is the star-delta method of starting. For this all six ends of the motor windings must be brought out to the starter.

On starting, the stator windings are first connected in star configuration. This is to reduce the voltage across each winding. This in turn, of course, reduces the current. When the motor has got up to speed the starter is switched so that the windings are in delta configuration with the full 415V across each winding. Diagrams 4.17 and 4.18 show how this can be carried out using a manual type star delta starter.

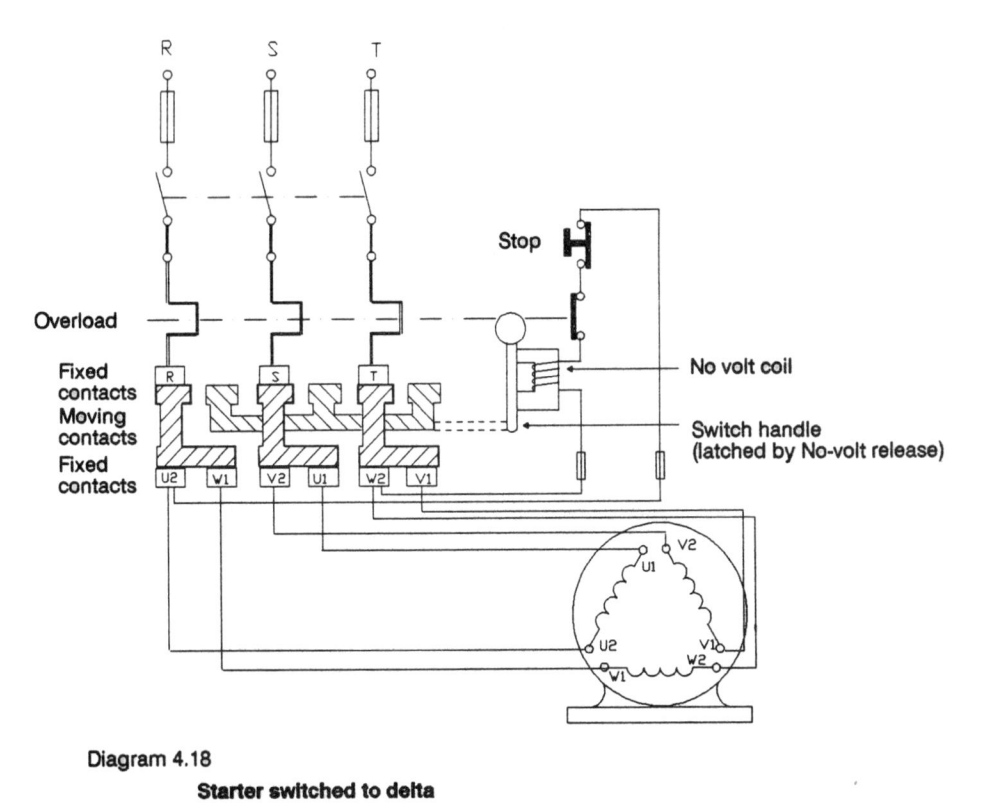

Diagram 4.17

Starter switched to star

Diagram 4.18

Starter switched to delta

40

Question 3

Draw the circuit diagram for a 415V three-phase motor started from a direct-on-line starter with a 240V coil. Label all parts.

Answer

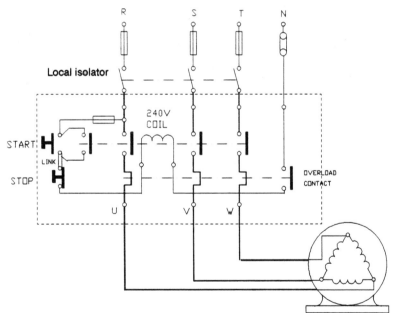

Diagram 4.19

Direct-on-line starter

Now try this - Three

a.   *Explain why overcurrent protection in motor starters must be designed so that a time delay is built in.*

b.   *Describe, with the aid of a labelled diagram, one example of how this can be achieved.*

Power of A.C. Motors

As a.c. motors are by design highly inductive it follows that this fact must be taken into account when calculating their load currents. A typical rating plate of a motor could include information such as kW output power, supply voltage and full load current. An example could show 4.8kW, 240V and 25A. The 25A could be used as the current when determining the overcurrent protection device for designing the circuit. It may pay however to look into it further. What is the power factor of the motor?

$$p.f. = \frac{watts}{volts \times amps} = \frac{4800}{240 \times 25} = 0.8$$

A value of between 0.8 and unity is usually acceptable but if it gets any lower it pays to correct this before carrying out any design of the circuit. If our example had a power factor of 0.5 this would be different. For a start the full load current would now be

$$power\ factor = \frac{watts}{volts \times amps}$$

$$amps = \frac{watts}{volts \times p.f.} = \frac{4800}{240 \times 0.5} = 40A$$

This shows an increase from 25A to 40A for the same output power of **4.8kW** and in this case it is well worth correcting the power factor. For our example we will correct it from 0.5 to unity.

So we have a power output of 4.8kW and VA of 240 x 40 = 9600 or 9.6kVA at the power factor of 0.5. The cos at 0.5 = 60°. With this information we can now construct a phasor diagram for power. Diagram 4.20

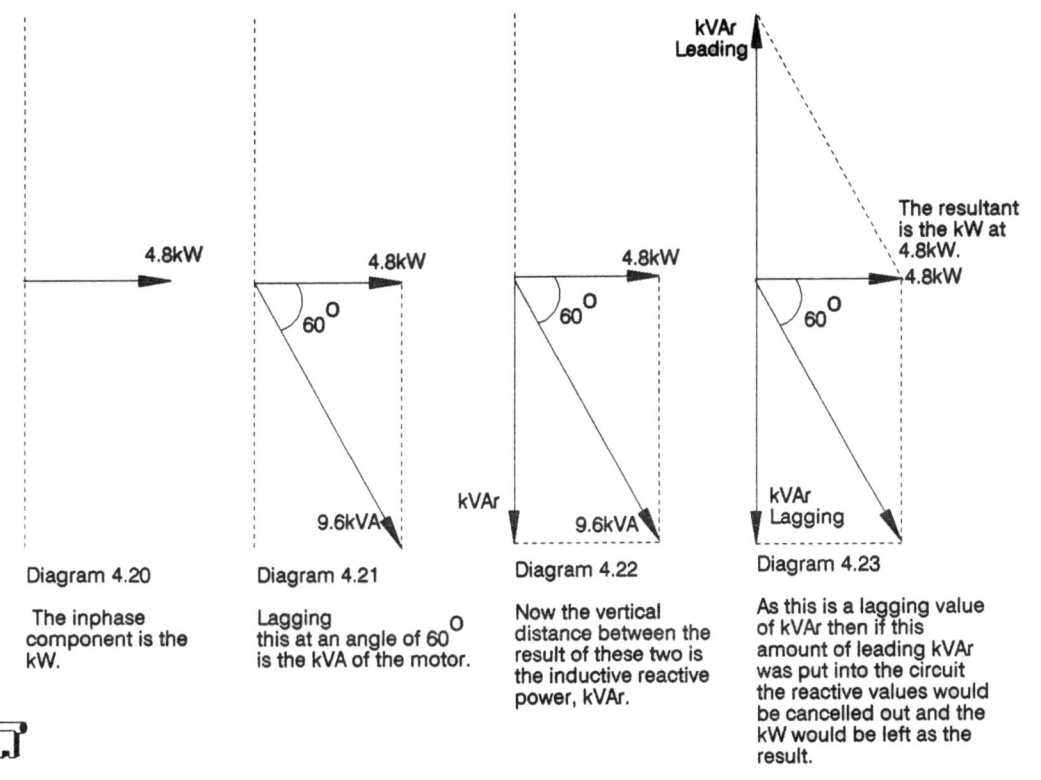

Diagram 4.20

The inphase component is the kW.

Diagram 4.21

Lagging this at an angle of 60° is the kVA of the motor.

Diagram 4.22

Now the vertical distance between the result of these two is the inductive reactive power, kVAr.

Diagram 4.23

As this is a lagging value of kVAr then if this amount of leading kVAr was put into the circuit the reactive values would be cancelled out and the kW would be left as the result.

By measuring the diagrams to scale the capacitor would need to produce a kVAr of 8.3 kVAr so that the motor that would effectively have a load current of

$$I = \frac{W}{U} = \frac{4800}{240} = 20A$$

So we have   at a p.f. of 0.5 load current   = 40A
at a p.f. of 0.8 load current   = 25A
at a p.f. of unity load current   = 20A

As you can see it is important to consider the power factor of a motor before calculating load currents and designing circuits.

Question 4

A motor has a voltmeter, ammeter and wattmeter connected to its supply cables. When running at full load the following readings were obtained:

    volts   - 238V
    amperes - 32A
    watts   - 5kW

a. Determine the power factor.
b. Determine the current that would flow through a capacitor circuit if the power factor was corrected to unity.

Answer

a. power factor $= \dfrac{W}{UA} = \dfrac{5000}{238 \times 32} = \dfrac{5000}{7616}$

   $= 0.65$

   the angle for cos 0.65 $= 49.5^{o}$

b. Drawing a power phasor diagram to scale

For the phasor diagram the leading kVAr = 5.77kVAr

The current related to the capacitor would be

   $I = \dfrac{P}{U} = \dfrac{5770}{238} = 24.24A$

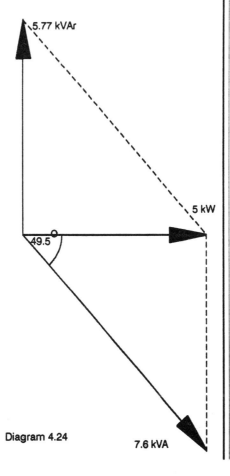

Diagram 4.24

What would the load current be in question 4 if the motor was corrected to a power factor of 0.8?

Load Current and Speed of A.C. Motors

All motors are inefficient, that is to say they have an efficiency of less than 100%. This, like the power factor, must always be taken into account when calculating the load current. One thing to remember is that after the efficiency and/or the power factor has been taken into account the load current will be greater than before.

Take an example where a single phase 2.8kW motor is only 75% efficient and the full load current is to be calculated. Without taking the efficiency into account the current would be

$$I = \frac{P}{U} = \frac{2800}{240} = 11.67A$$

Remember that after the efficiency has been taken into account the current will be greater than this.

$$75\% \text{ is } \frac{75}{100} = 0.75$$

$$75\% \text{ efficient related to } 11.67A = \frac{11.67}{0.75} = 15.56A$$

If this motor had a power factor of 0.6 the full load current would now be

$$I = \frac{15.56}{0.6} = 25.9A$$

Another factor that must be considered is the speed of the motor when related to the frequency and number of poles. If an a.c. motor working on 50Hz (50 cycles per second) had one pair of poles it would have a theoretical speed of 50 revolutions in one second or 50 x 60 = 3000 revolutions in one minute.

 Poles should always be taken in pairs. Remember a magnet must have a north and a south. A four pole machine has two pairs, and when working on 50Hz will complete one revolution every 2 cycles so the speed will be:

$$\frac{50Hz \times 60 \text{ seconds}}{2 \text{ pairs of poles}} = 1500 \text{ revs/min}$$

Half the speed of a two pole machine.

The speed calculated here is not the speed of the shaft but the speed of the rotating magnetic field. This is known as the synchronous speed. The shaft speed will always be less than this. The difference between the shaft speed and the synchronous speed is known as the "slip speed".

This can be calculated from

$$s = \frac{N_S - N_r}{N_S}$$

where
s = slip
N_S = synchronous speed
N_r = rotor speed

This can be calculated as a percentage; in which case

$$(s)\% = \frac{N_S - N_r}{N_S} \times 100$$

Question 5

a. *A 415V three-phase motor is found to have an efficiency of 85% on full load. The power factor at this time is 0.8 lagging when the full load is 5.6kW. Calculate the current of the motor under these conditions.*

b. *If an eight pole induction motor runs at 12 rev/sec and is supplied from a 50Hz supply calculate the percentage slip.*

Answer

a. The power in a three-phase motor $= \sqrt{3}UI \times$ power factor

$$I = \frac{P}{\sqrt{3}\ UI \times \text{p.f.}} = \frac{7000}{\sqrt{3} \times 415 \times 0.8}$$

$$= 12.18A$$

At an efficiency of 87%

$$I = \frac{12.18}{0.87} = 14A$$

b. An eight pole machine on 50Hz has a synchronous speed of

$$Ns = \frac{50 \times 60}{4} = 750 \text{ rev/min}$$

and a rotor speed of

$$Nr = 12 \text{ rev/min} = 12 \times 60 = 720 \text{ rev/min}$$

$$\text{Percentage slip} = \frac{Ns - Nr}{Ns} \times 100 = \frac{750 - 720}{750} \times 100 = 4\%$$

Now try this - Five

When connected to a 240V 50Hz supply a 7.5kW motor takes a current of 55A with a power factor of 0.8 lagging. Calculate the efficiency of the motor.

Tips to help you answer the "Now try this" questions in this section.

Now try this - One

Remember the sequence from the supply is always Red, Yellow, Blue. The motor will follow this sequence.

Now try this - Two

The materials that make up the electrical and magnetic circuits are either good electrical conductors such as copper or aluminium, or good magnetic conductors such as laminated iron. Other materials make up the mechanical parts for rotation, cooling and the case.

Now try this - Three

Remember that when a motor starts from a stationary position a large current is required.
The overload devices use either the thermal or magnetic effects of current flow.

Now try this - Four

By redrawing the phasor diagram in Diagram 4.24 it is possible to add a line at an angle of cos 0.8. A new kVAr can now be plotted and a resultant of the difference between that of unity and 0.8 be determined. It only remains then to calculate the current from the resultant kVAr.

Now try this - Five

Remember that the efficiency of a motor is: $\dfrac{\text{output}}{\text{input}} \times 100$

where it is given as a percentage. The output is the power in watts developed by the motor. The input will also be in watts but will always be greater than the output.

Further Revision

1. Explain, with the aid of diagrams, how the following motors can have their direction of rotation reversed:

   i)   universal (series motor)
   ii) three-phase wound rotor motor

2. Describe how the windings of a totally enclosed motor are cooled.

3. Some motors have thermistors embedded in their stator windings. Explain how these can prevent a motor from being damaged due to overheating.

4. When working on full load a 240V, 50Hz single-phase induction motor has a power of 1.75kW. If the power factor of the motor is 0.7 what is the input current?

5. The resistance of the windings of a 240V single-phase motor is 9Ω with an inductive reactance of 12Ω. A capacitor with an Xc of 24Ω is connected across the motor for power factor correction. Calculate:

   i)   the impedance of the motor winding
   ii)   the current flowing through the motor winding
   iii) the current flowing through the capacitor
   iv)   the total current taken from the supply

6. A motor has a full load power of 10kW when connected to a 415V three-phase supply. If the motor efficiency is 75% calculate the current flowing in its supply cables.

7. A washing machine motor has two sets of windings within it so that two speeds can be obtained. For the washing action the motor is connected as a 4 pole machine and has a rotor speed of 1450 rev/min. When it is being used for spin drying the motor is used as a 2 pole machine with a rotor speed of 2800 rev/min. If it is connected to a 240V 50Hz supply what are the percentage slips at each speed?

8. Using a fully labelled block diagram show the operation of a washing machine control system. Include water level sensors and water temperature sensors feeding information to a controller timer unit which operates a pump, heater and drum motor.

Section Five

Lighting

These questions cover the area of lighting sources and control.

1.a. Compare the relative efficacies of EACH of the following types of lamp:

    i)   low pressure sodium vapour
    ii) high pressure mercury vapour
    iii) tungsten halogen

b. Compare the low pressure sodium vapour lamp to the tungsten halogen with regard to colour rendering.

c. Give applications for

    i)   Low pressure sodium lighting
    ii) Tungsten halogen lighting

2.a. Draw the circuit diagram for a switch start fluorescent luminaire.

b. Explain the function of each component shown in the circuit for (a).

3. Diagram 5.1 shows the circuit diagram for a high pressure mercury vapour lamp.

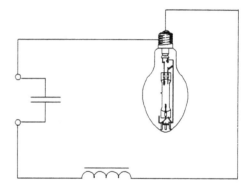

Diagram 5.1

a. Describe the starting process for this lamp.
b. Give TWO applications for this type of lamp.

4. Compile two lists, one giving lamps that may cause the stroboscopic effect on rotating machinery, the other giving methods of reducing this effect.

5. Explain how semiconductor devices can be used to control the light output of a tungsten lamp, without high losses in the control equipment. Use a waveform drawing to explain the operation.

Introduction

There are a number of different types of lamps in use and it is important to have some knowledge of each of them. For a start how can you compare one with another? How are the circuits made up and what are the applications?

In addition to this it is useful to know how lamps can be controlled to give dimming effects without damaging the equipment.

Comparison of Light Sources

To compare the relative inputs and outputs of lamps, their efficacies are 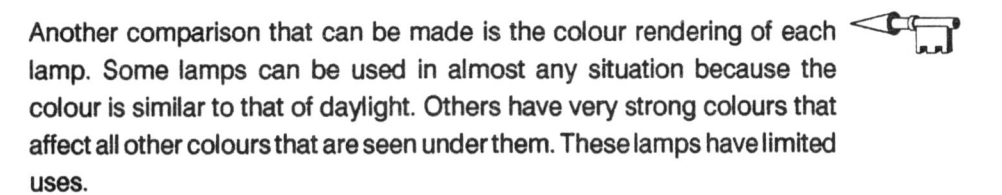 used. Efficacy is measured in lumens per watt and it is the value calculated from the power input to a lamp to the light output. This automatically takes into account power lost in control gear so it is a useful comparison. It should not get confused with efficiency which measures power in to power out.

Another comparison that can be made is the colour rendering of each lamp. Some lamps can be used in almost any situation because the colour is similar to that of daylight. Others have very strong colours that affect all other colours that are seen under them. These lamps have limited uses.

Examples of Lamp Efficacies	
Type of lamp	**Efficacy (lm/Watt)**
(GLS) Tungsten Filament	10 - 18
(TH) Tungsten Halogen	12 - 22
(MPF) HP Mercury	32 - 58
(MBT/F) Mercury Tungsten	10 - 26
(MBF/U) Mercury Fluorescent	36 - 54
(MB1F) Metal Halide	66 - 84
(MCF) Fluorescent	37 - 90
(CFL) Compact Fluorescent	65 - 71
(SON) HP Sodium	55 - 120
(SOX) Low Pressure Sodium	70 - 160

Question 1

a. Compare the relative efficacies of EACH of the following types of lamp:

    i) *low pressure sodium vapour*

    ii) *high pressure mercury vapour*

    iii) *tungsten halogen*

b. Compare the low pressure sodium vapour lamp to the tungsten halogen with regard to colour rendering.

c. Give applications for

    i) *Low pressure sodium lighting*

    ii) *Tungsten halogen lighting*

Answer

a.   Typical efficacies for each lamp are as follows

    i)   Low pressure sodium 70-160

    ii) High pressure mercury vapour 32 - 58

    iii) Tungsten Halogen 12-22

    The highest efficacy is the Low pressure sodium, then the High Pressure mercury vapour and the Tungsten halogen.

b.   The light output of a LP sodium lamp is monochromatic. This means that it only uses one colour from the light spectrum. The colour it gives out is yellow. Any other colour coming under it is affected by it. Colours such as red show brown, blue becomes green and white shows as yellow.

    Tungsten halogen lighting does not have this effect on colours, in fact it tends to show bright colours up even better.

c.   i) The effect that LP sodium lighting has on colours means its applications tend to be limited to street lighting and outside security lighting. Its high efficacy means that it is very cost effective to use where outside lights have to be left on for long periods.

    ii) Tungsten halogen lighting has many uses both indoors and outdoors. The fact that it requires no control gear makes it better suited to some applications where discharge lighting may otherwise be used.

    Tungsten halogen is ideal for display lighting where a sparkle is required, such as for jewellery.

Now try this - One

What precautions should be taken when replacing a tungsten halogen lamp? (include checks for isolation)

Fluorescent Luminaires

The fluorescent tube is by far the most common type of discharge lamp in use. It is a low pressure mercury vapour lamp which requires the cathodes being heated up at each end of the lamp before it will work. It also requires a high voltage between the cathodes to make it "strike" and create the discharge.

As we have seen in other sections of this revision book the inductive effects of coils when used on a.c. is very important. The choke or ballast 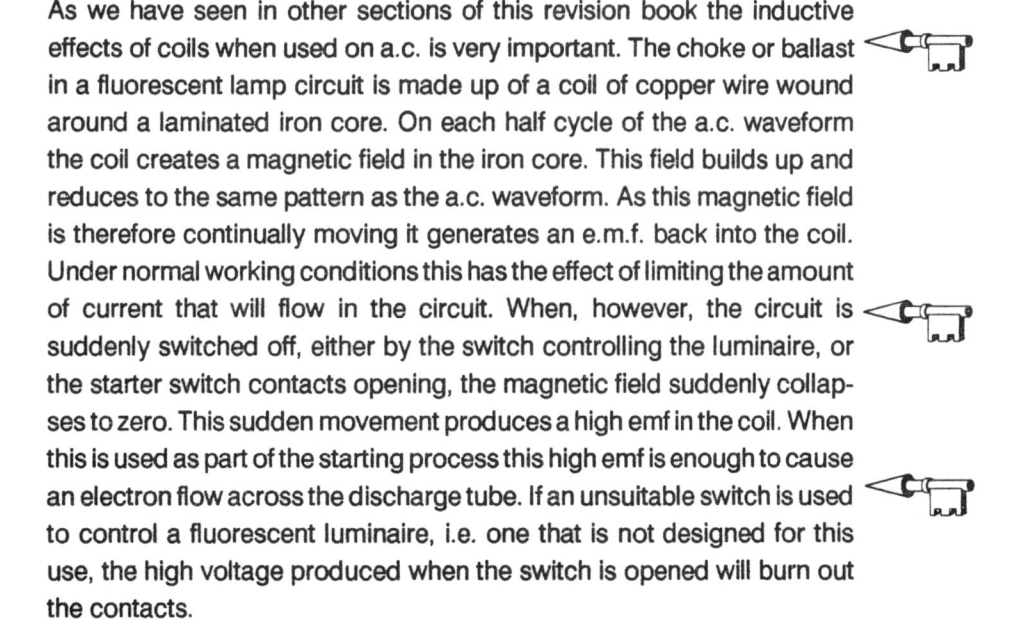 in a fluorescent lamp circuit is made up of a coil of copper wire wound around a laminated iron core. On each half cycle of the a.c. waveform the coil creates a magnetic field in the iron core. This field builds up and reduces to the same pattern as the a.c. waveform. As this magnetic field is therefore continually moving it generates an e.m.f. back into the coil. Under normal working conditions this has the effect of limiting the amount of current that will flow in the circuit. When, however, the circuit is suddenly switched off, either by the switch controlling the luminaire, or the starter switch contacts opening, the magnetic field suddenly collapses to zero. This sudden movement produces a high emf in the coil. When this is used as part of the starting process this high emf is enough to cause an electron flow across the discharge tube. If an unsuitable switch is used to control a fluorescent luminaire, i.e. one that is not designed for this use, the high voltage produced when the switch is opened will burn out the contacts.

It depends on the particular circuit and control equipment used as to how the discharge is started and controlled. The most used circuit consists of a series choke (ballast) and a switch starter. To overcome the inductive effects of the choke a power factor improvement capacitor is connected across the supply.

Question 2

a. *Draw the circuit diagram for a switch start fluorescent luminaire.*

b. *Explain the function of each component shown in the circuit for (a).*

Answer

a.

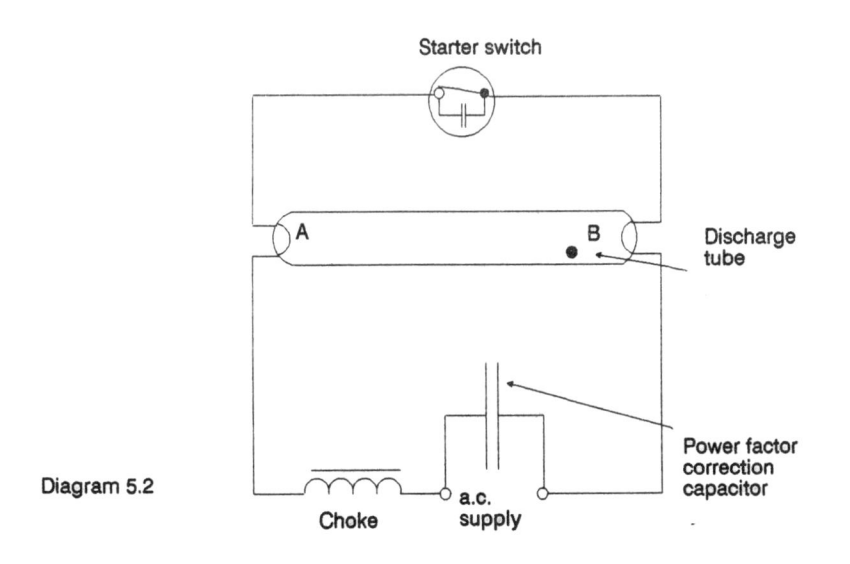

Diagram 5.2

b. When the supply is connected to the circuit, current flows through the choke, through the lamp filament A, the starter, the second lamp filament B and back to the supply. The gas in the starter heats up and the bimetal contacts close. The gas now cools down and the contacts open. During this time the two filaments in the tube have heated up and an electron cloud has been created around them. The opening of the contacts in the starter stops the current flow causing a collapse of magnetic field in the choke, inducing a high voltage. This voltage strikes the tube and starts the electron flow through the low pressure mercury vapour. The low resistance of the tube now shorts out the starter switch. Once the tube has a discharge flowing through it, the choke becomes a current limiting device.

If the lamp is switched off and then back on again the procedure is repeated and the lamp will strike almost immediately.

The power factor correction capacitor brings the current back into line with the voltage so that the current in the circuit conductor is kept to a minimum.

Now try this - Two

If the power factor correction capacitor is removed from a discharge lighting circuit what will the effect be on:

    i)   the starting of the lamp

    ii) the current flowing in the luminaire

    iii) the current flowing in the circuit supplying the luminaire

High Pressure Mercury Vapour Lamps

The high pressure mercury vapour lamp consists of an arc tube inside an outer glass envelope.

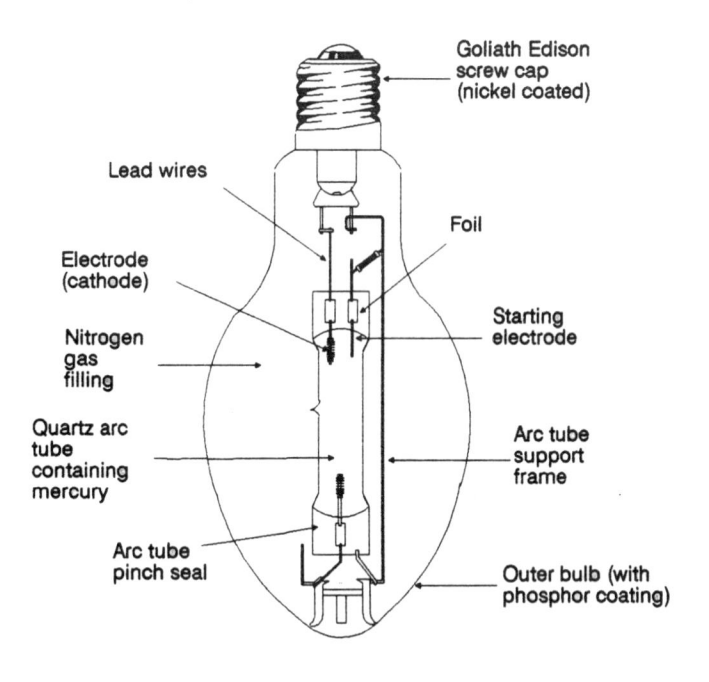

- Goliath Edison screw cap (nickel coated)
- Lead wires
- Foil
- Electrode (cathode)
- Starting electrode
- Nitrogen gas filling
- Quartz arc tube containing mercury
- Arc tube support frame
- Arc tube pinch seal
- Outer bulb (with phosphor coating)

Diagram 5.3

As the electrodes are not pre-heated in this type of lamp their construction is solid rather than coiled wire. A material that emits electrons when heated is built into the electrode.

The starting of the HPMV lamp is assisted by the auxiliary electrode placed very close to the main electrode which is closest to the cap.

There are many applications for this type of lamp but it must be remembered that if it is switched off even for a second then it may take up to five minutes before the pressure can fall and the lamp light again. This can limit the applications in some cases.

Question 3

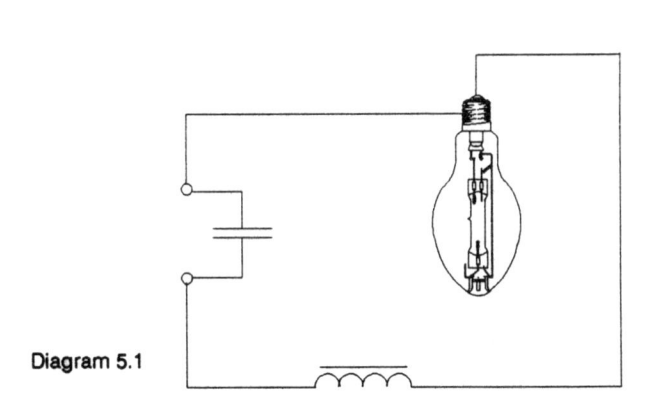

Diagram 5.1

Diagram 5.1 shows the circuit diagram for a high pressure mercury vapour lamp.

a. Describe the starting process for this lamp.
b. Give TWO applications for this type of lamp.

Answer

a. Inside the discharge tube of a high pressure mercury vapour lamp there are two main electrodes and one auxiliary electrode. The auxiliary is placed close to the main electrode closest to the cap and has a resistor connected as shown in Diagram 5.4

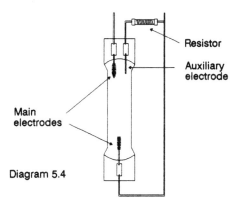

Resistor

Auxiliary electrode

Main electrodes

Diagram 5.4

When the circuit is first switched on a small discharge, in the form of an arc, starts between the main electrode and the auxiliary. This raises the temperature of the emissive material in the main electrode. The current from the discharge passes through the auxiliary resistance causing a voltage drop. The starting arc then collapses creating a high voltage from the induced e.m.f. in the choke. This now strikes the electron flow across the mercury vapour between the two main electrodes. The lamp then builds up pressure until it is up to full light. As with low pressure mercury circuits the choke becomes a current limiting device once the arc has struck.

b. It is used a great deal for street lighting and can also be found in sports halls where they are mounted at high level.

Now try this - Three

Explain why, when the high pressure mercury vapour lamp is switched off, even for a few seconds, it takes time to get back to full brilliance.

Stroboscopic Effect

When an a.c. supply is connected to a discharge circuit the electron flow  across the lamp changes direction every time the a.c. cycle goes onto the other half. This means that at the point of change-over there is no electron flow and the lamp is off.

As the effect is produced by a flashing light it is the "off" effect of the flash that must be reduced. A simple way of achieving this is to place two fluorescent tubes side by side connected so when one tube is off the other is still glowing. Fluorescent luminaires are made with twin lamps especially for this. One lamp is wired in a conventional way, while the other has a capacitor in series with it causing a leading phase shift.

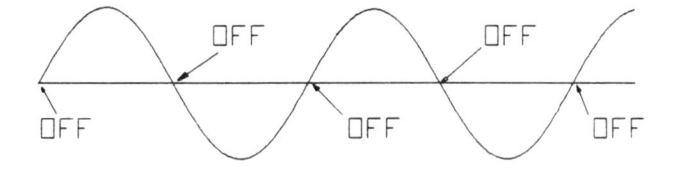

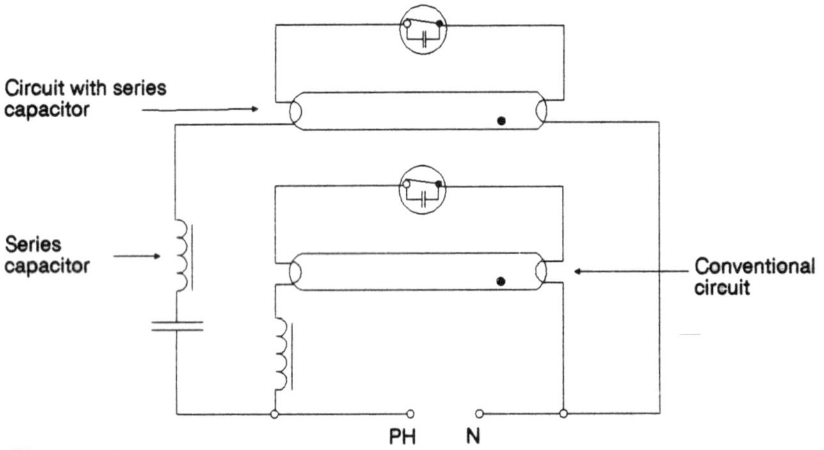

Diagram 5.5

On a 50Hz supply this is difficult to see with the naked eye. Machines that rotate can appear to be affected by this though. A rotating shaft under fluorescent lighting can appear to be going slow, backwards or even be stationary when, in fact, it is still spinning at thousands of revolutions per minute. The effect can, of course, be very dangerous and should be reduced or eliminated whenever possible.

Diagram 5.6

Lead-lag circuit

Where there is a three-phase supply available rows of single tube luminaires can be used and connected to different phases. As the three phases reach the zero point in their waveforms 120° apart the lamps have their "off" times at different stages and the stroboscopic effect is reduced.

As machine manufacturers do not necessarily know their customers' lighting arrangements, low voltage tungsten halogen luminaires are often prefitted to machines. The bright white light, which is continuous, overrides the stroboscopic effect of fluorescent luminaires positioned further away.

Question 4

Compile two lists, one giving lamps that may cause the stroboscopic effect on rotating machinery, the other giving methods of reducing this effect.

Answer

Those that cause stroboscopic effect	Those that can reduce stroboscopic effect
single fluorescent luminaires (on 50Hz supply)	lead lag fluorescent luminaires
HP mercury vapour lamps	fluorescent luminaires on different phases
LP sodium lamps	high frequency fluorescent luminaires
HP sodium lamps	tungsten filament lighting
	tungsten halogen lighting

Now try this - Four

A small repair workshop with a single-phase supply has a drilling machine and small lathe. The general lighting is by single fluorescent luminaries. The rotating machines are occasionally affected by the stroboscopic effect. Suggest how this problem may be overcome, and give reasons for your choice.

Light Dimmer Circuits

The electronic dimmer circuit should not be confused with trying to reduce light by using resistors. The method using resistors reduces the voltage across the lamp and creates unwanted heat. The size of resistors required for this make it impractical.

The electronic dimmer uses a method of reducing power not voltage. The  theory can be explained by comparing the action of the electronic circuit to that of a light switch. If a filament lamp is rapidly switched on and off the filament will not completely heat up or cool down but will glow somewhere below full brilliance. As no current flows when the switch is off the power consumed is less than maximum. Obviously it is impractical to manually switch a light in this way, so an electronic dimmer circuit is used instead. The control on the dimmer increases or decreases the switching speed of the lamp. To increase the brightness of the lamp the "on" periods are made longer than the "off".

The electronic circuit uses the natural switching effect of the a.c. waveform. Remember that on a 50Hz supply, as the waveform goes through zero the supply is off 100 times every second. The device used in the electronic circuit to control the lamp is a "triac". This is made so that it  will conduct when switched on and stop when the a.c. waveform reaches zero. If it is switched on at the beginning of a half cycle it will conduct until the next zero point. Diagram 5.7

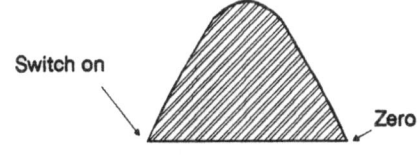

Diagram 5.7

If it is switched on near the end of the wave it will only conduct a short time before the zero.
Diagram 5.8

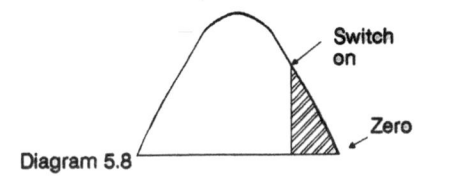

Diagram 5.8

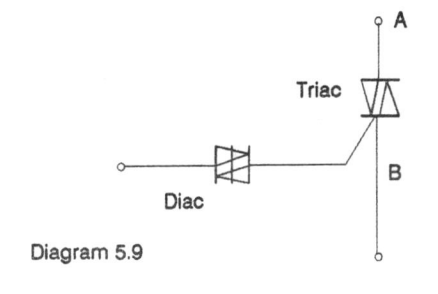

Diagram 5.9

The device that "fires" the triac into action and starts it conducting is a diac. When the diac switches the triac on current can flow through the main circuit A - B in Diagram 5.9. As soon as the wave reaches zero current stops flowing until the diac switches the triac on.

So that the timing of the diac is controlled a variable resistor and capacitor are incorporated.
Diagram 5.10
As the variable resistor is adjusted a charge builds up on the capacitor and at a pre-set voltage the diac triggers the triac to conduct. The current flowing through the variable resistor is very small, so little or no heat is produced. By varying the resistor value the time can be varied as to what part of the waveform the triac is switched on.

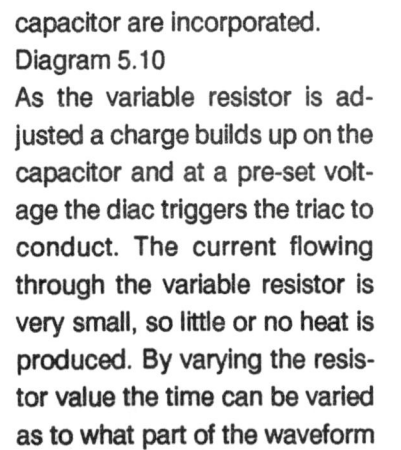

Diagram 5.10

As this is a switching circuit it can cause some radio interference. To overcome this a suppression circuit consisting of a resistor and capacitor is added to the basic dimmer circuit.

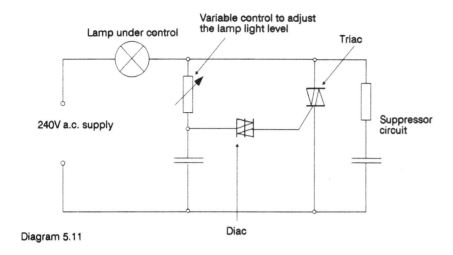

Diagram 5.11

Question 5

Explain how semiconductor devices can be used to control the light output of a tungsten lamp, without high losses in the control equipment. Use a waveform drawing to explain the operation.

Answer

An electronic dimmer uses the a.c. waveform to switch the tungsten lamps continually ON and OFF. A diac is used to start a triac conducting. This will continue to conduct until it is switched off. As the a.c. waveform switches off every half cycle the triac will conduct until the end of that particular half cycle.

The earlier the half cycle is switched on the greater the light output. If the cycle is not switched on until very late in its wave only a small amount of power is allowed to be used and the light level is low.

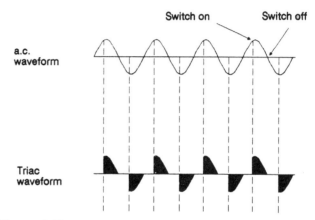

Diagram 5.12

Now try this - Five

An electronic dimmer switch is to be installed in replacement for a two-way lighting switch. List the factors that must be considered before this can be carried out.

Tips to help you answer the "Now try this" questions in this section.

Now try this - One
Remember the tungsten halogen lamp uses a quartz envelope.

Now try this - Two
Remember that the power factor correction capacitor is there to reduce the inductive effect of the choke.

Now try this - Three
Remember that when a high pressure mercury vapour lamp is working at full brilliance the pressure inside the inner bulb is high at high temperature. Remember also that the starting procedure takes place when the pressure is low.

Now try this - Four
Look at the facts you are given, there is no three-phase so that method cannot be used. You are therefore limited to what is available. Don't forget to give consideration to cost implications when you make your choice.

Now try this - Five
Remember that the switch action is independent to the electronic dimmer circuit.

Further Revision

1.a. Draw a labelled circuit diagram for a 150W low pressure sodium lamp.

b. Describe the starting process for this type of lamp.

2.a. Explain why the lamp ratings of fluorescent tubes cannot be taken when calculating the load current of a circuit.

b. What is the minimum number of circuits that can be used with the loads shown below if 5A protection devices are used?

15 twin lamp 60W fluorescent luminaires
24 single lamp 70W fluorescent luminaires

3.a. Explain what is meant by stroboscopic effect when related to discharge lighting.

b. Describe what will happen if the "glow" type starter is removed from a fluorescent luminaire when the tube is on.

4. Explain why an electronic dimmer designed for resistive loads should NOT be used on fluorescent luminaires.

5. List the factors that should be considered when disposing of the following unwanted discharge lamps:

i) fluorescent
ii) high pressure mercury
iii) low pressure sodium

Section 6

Inspection and Testing

This section considers the answers to these questions.

1. a. Explain why it is important to carry out an inspection of an installation before any tests are carried out.

   b. List FOUR pieces of documentation required before an inspection and test can be carried out.

2. State the instruments required, and their ranges, for the following tests:

   a. continuity of protective conductors
   b. continuity of ring final circuits
   c. insulation resistance
   d. polarity
   e. earth fault loop impedance

3. With the aid of diagrams explain how to carry out an insulation resistance test on a final circuit from a single-phase distribution board. It is known that at least one circuit contains an electronic dimmer switch.

4. The working voltage and current of each phase is to be taken at a three-phase induction motor.

   The three-phase cage induction motor is installed at the end of a long cable run. Checks have to be carried out to ensure the calculations for voltage drop and load current conform to the actual equipment when in use. Describing all necessary safety precautions explain how the

   a. voltage can be measured
   b. current in each phase can be determined when the motor is working on full load

5. A bank of fluorescent luminaires appears to be drawing more current than it was designed for. An ammeter reading indicates that a current of 8A is flowing and the circuit voltage is 240V at 50Hz. A wattmeter reading shows the power is 960W. The customer wants to know (a) what the power factor is and (b) what the current would be if the bank of fluorescent luminaires was corrected to unity power factor.

Introduction

How do we know if there is a fault somewhere in an installation?

We can, of course, connect up the supply and see what happens. A fuse may blow, a piece of equipment start smoking, you or your colleagues may get an electric shock. Nothing may happen, it could be months or even years before something causes a fault to show up. The thing is that unless an installation is inspected and tested properly - nobody knows.

In addition to the inspection and testing that must be carried out on the completion of work there are other times meter readings may need to be taken. At all times safety must be given consideration when carrying out tests. This not only applies to those personnel carrying out the test but also to anyone else who may come into contact with it.

Inspection

The inspection of an installation should be an ongoing process while the  wiring systems are being installed. This can often save time in the future when much of the wiring is concealed. There are certain things that must be inspected at different stages in the installation process. For example all conduit runs must be complete before any cables are installed and to confirm they are complete and all connections are tight, an inspection must be carried out.

An inspection can often reveal odd things that have been overlooked and could be dangerous if left. The equipotential bonding conductor that was installed but couldn't be connected at the time because the room was not complete is an example. Without an inspection this could be missed and some item left unbonded. Inspection does not stop at just looking at completed work, it may involve removing accessory plates and looking inside, checking connections are tight, all mechanical protection is in place, cord grips are used correctly......... Removing a ceiling rose cover can reveal a multitude of sins!

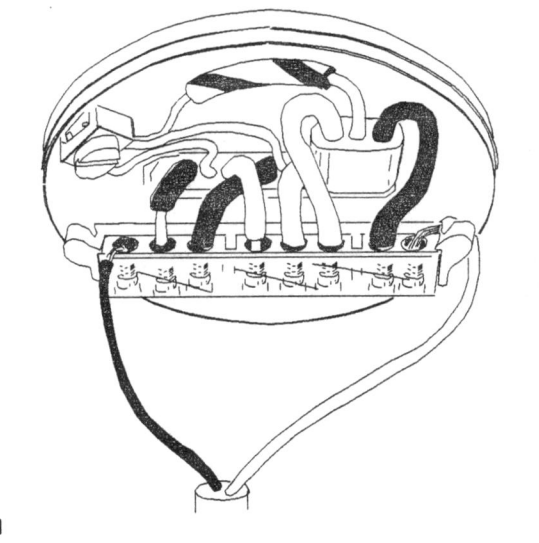

Diagram 6.1

Question 1

a. *Explain why it is important to carry out an inspection of an installation before any tests are carried out.*

b. *List FOUR pieces of documentation required before an inspection and test can be carried out.*

Answer

a. If testing takes place without first carrying out an inspection then time and money could be wasted and dangerous situations could arise. Most of the tests use a power source of some type. If an inspection has not been carried out cables may not have been connected and could be left hanging around. When the test voltages are applied these cables could become a dangerous hazard and cause an accident.

It is also important to ensure circuits and equipment are switched correctly before any test instrument is connected. Test voltages can cause damage to equipment as well as cause accidents.

b. Regulation 514-09-01 outlines the documentation required and includes:

i) Installation specification
ii) Plans and drawings
iii) Design information, showing:
   - circuit arrangements - conductors and protection devices
   - any equipment vulnerable to tests
   - isolation and switching
iv) Charts for recording results and comparing with design data.

Now try this - One

The fixing screws on a 13A socket outlet are removed so that an inspection can be carried out. List the points that should be considered.

Test Instruments

In Part 6 of the Wiring Regulations and 9.3.1 of the On Site Guide a list is shown of the tests that may need to be carried out. It is important that correct equipment is used for each test otherwise the results may be misleading.

Continuity of protective conductors - we shall assume that these are copper conductors either as part of a cable, such as p.v.c. twin and cpc, or a separate single insulated cable. The resistance of these conductors can be very low especially where they have a large cross sectional area. Resistances of less than 1Ω may need to be recorded so a meter capable of this should be used. As this is is a continuity test and it is checking the cable is continuous throughout its length, a high voltage is not required. Voltages of between 6 and 9 volts are typical.

If the circuit protective conductor is made of steel conduit then a separate high current test may need to be used.

Continuity of ring final circuits - is another continuity test that will be 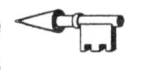 recording low resistance values. Where the ring circuit being tested is short the resistance may be below 0.01Ω.

Insulation resistance tests - unlike the two above are measuring the  insulation resistance not the conductor resistance. It is important to ensure that the conductors are not coming together at any part of their length. It is also important to test that the insulation is not likely to break down under pressure. The way of applying this pressure test is to apply a high voltage across the insulation. The test voltage for installations supplied with 240V or 415V is 500V d.c. at a current of 1mA. As it is insulation that is being measured the results should be in the millions of ohms range.

Polarity - is a test that checks on the correct wiring of circuits and no value is obtained that should be recorded. A continuity meter may be used for this test but often a bell or buzzer is used from a low voltage battery supply.

Earth fault loop impedance tests - are carried out after the supply has been connected. The test instrument injects a current of up to 25A through the earth fault path and records the resultant value in ohms.

Earth electrode resistance - this is a specialist test which may require special test equipment. However, if the earth electrode to be tested is part of a TT system, an earth fault loop impedance tester may be used.

Operation of residual current devices - requires equipment especially designed for the test. The test equipment injects small currents through the rcd and checks if it trips or not. If it does trip the time it took must be measured.

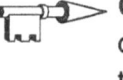

Question 2

State the instruments required, and their ranges, for the following tests:

    a. *continuity of protective conductors*
    b. *continuity of ring final circuits*
    c. *insulation resistance*
    d. *polarity*
    e. *earth fault loop impedance*

Answer

The instruments required for the tests are as follows:

a.   Continuity of protective conductors - an ohmmeter with a range starting at less than 1 ohm.

b.   Continuity of ring final circuits - an ohmmeter with a range starting at less than 0.1 ohm.

c.   Insulation resistance - an ohmmeter that uses a test voltage of 500V d.c. at a current of 1mA and measures resistance values up to $1M\Omega$.

d.   Polarity - an instrument similar to that used for continuity testing or a bell set that will operate on resistances up to 1 ohm.

e.   Earth fault loop impedance - this instrument must be calibrated in ohms but works on currents up to 25A. The resistance range must be capable of readings down to 0.1 of an ohm.

Now try this - One

Explain what safety precautions must be taken before any tests are carried out?

Insulation Resistance Testing

All testing should be carried out in such a way that safety is always given  a top priority. When insulation resistance tests are to be carried out it must be remembered that a voltage of 500V d.c. is used. Although this is not a lethal voltage it can create serious situations. For example someone standing on steps putting up a luminaire could receive a shock from the test voltage and fall.

Insulation resistance tests are often carried out on extensions to installations or are part of a periodic inspection and test. In these situations live supplies are present. Before an insulation resistance test can be carried out checks should be made to ensure no live conductors are accessible. An isolation test procedure should be adopted before any insulation resistance test instrument is connected.

Many insulation resistance test instruments now use batteries as their source of power. If the supply of these is low, incorrect readings will be given. Carry out a battery check before any recordings are made.

After a time test instruments become inaccurate and it is important that  calibration is carried out at regular intervals.

Over the years many insulation resistance testers have been developed but the specifications that were used previously do not necessarily meet the accepted requirements now. It is always important to check that the instruments used are "up-to-date" and the readings they give are valid.

So that no confusion can exist as to what instrument has been used for particular tests the serial number of the equipment should always be recorded on the results sheet.

Before any tests are carried out the condition of the test leads should be checked for damage. Tests should be carried out with the leads open circuit and short circuited to ensure the correct readings are obtainable.

An insulation resistance test is carried out to confirm there is no connection between the live conductors or the live conductors and earth. If equipment is in the circuit and connected between phase and neutral or between phases the test voltage will also be put across this. Apart from the fact false readings will be given, serious damage may be caused to the equipment. Before testing between phase and neutral ALL equipment, lamps, electronic devices and portable appliances must be disconnected. Electronic dimmer switches must be shorted out so that they are not damaged and all of the circuit can still be tested.

When all of this has been carried out testing can be considered.

Question 3

With the aid of diagrams explain how to carry out an insulation resistance test on a final circuit from a single-phase distribution board. It is known that at least one circuit contains an electronic dimmer switch.

Answer

The first consideration when carrying out an insulation resistance test from a distribution board must be safety. Tests must be carried out to ensure that the final circuit to be tested is "dead" and there are no exposed live conductors that could be touched when carrying out the test. Once it has been confirmed the circuit is safe to work on an inspection of the circuit should be carried out. All lamps should have been removed and switches put in the "on" position. Fluorescent luminaires must have been disconnected or unplugged. Dimmer switches should have been shorted out and a link put across two-way switches. When it is clear that all of this is complete the test can be carried out.

First check the meter for battery condition and see that the leads and connections are not damaged. Short the leads together and test the instrument. If everything is working correctly, connect one lead to the circuit neutral and the other to the phase conductor. Diagram 6.2

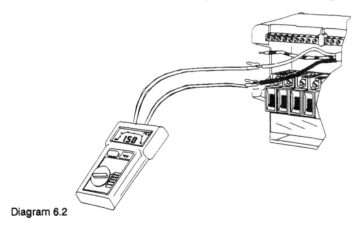

Diagram 6.2

When the reading is taken the resistance should exceed 0.5MΩ. With a short lead connect the phase and neutral together and test between this link and the circuit protective conductor.

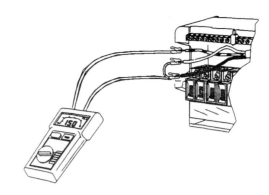

Diagram 6.3

This reading should also be greater than 0.5MΩ. When the readings are complete the circuit should be restored to its original condition, i.e. links removed, lamp replaced and equipment reconnected.

Now try this - Three

Explain how you would carry out an insulation resistance test on a three-phase delta connected induction motor.

Voltage and Current Readings

It is sometimes necessary to take voltage and current readings on circuits when the supply is connected. These should always be carried out with care and in such a way that the possibility of getting a shock is negligible. To do this the equipment must be in good condition and comply to all safety recommendations.

Voltage Readings

Voltage readings should only be taken if everything is shrouded with  insulation to such an extent that nothing live could be touched with your fingers. The test probes should conform to the requirements of the Health and Safety Executive Guidance Note GS38. An example is shown in Diagram 6.4.

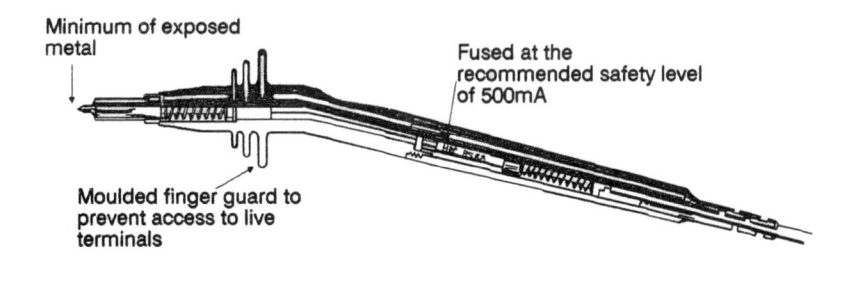

Minimum of exposed metal

Fused at the recommended safety level of 500mA

Moulded finger guard to prevent access to live terminals

Diagram 6.4

Test Probe

Reproduced with kind permission of Robin Electronics Ltd., Wembley.

Under these conditions only the very tip of the test probe can touch the live conductors and the flexible insulated leads with shrouded ends ensure the voltages to the test instrument are safe. Under these conditions voltages can be taken, with care, with very little or no risk of shock.

Current Readings

To use a conventional ammeter to take current readings, the circuit has to be broken and the ammeter connected in series with the load. Using a clamp-on type ammeter no circuit connections have to be connected or disconnected. The only requirement is that it is possible to get at a single circuit conductor. This conductor must be insulated. The ammeter can be clamped around this and the reading taken in amperes. This instrument should only be used on single insulated conductors where there are no exposed live parts.

Diagram 6.5

Clamp-on ammeter

Question 4

The working voltage and current of each phase is to be taken at a three-phase induction motor.

A three-phase cage induction motor is installed at the end of a long cable run. Checks have to be carried out to ensure the calculations for voltage drop and load current conform to the actual equipment when in use. Describing all necessary safety precautions explain how the

   a. *voltage can be measured*
   b. *current in each phase can be determined when the motor is working on full load*

Answer

a. When measuring the voltage at an a.c. three-phase motor care must be taken first to ensure there is no chance of injury due to the mechanical movement of the motor drive mechanism or the load. Next the correct instrument must be selected. This would be a meter set to a.c. voltage and capable of readings up to and including 415V. The leads and connections should be checked to ensure the insulation is not damaged. The test prods should be of the appropriate type.

With the motor isolated the terminal cover should be removed. If necessary conductors should be adjusted so that a clear path and sight is available to each connection. It may be advisable to try the test probes on each terminal while the motor is still isolated. If there is any chance of personal contact with terminals while carrying out the tests then extra insulation may need to be put in place. When satisfied that the tests can be carried out safely the motor should be started up. The voltages between each of the phases should be measured, recorded and compared with the design calculations. The terminal cover should be replaced before leaving.

b. As the load current of the motor will be the same throughout the length of cable run, the current does not have to be measured at the motor terminals. A clamp-on ammeter would be used as this does not require breaking into the circuit. Any point in the cable run where individual phase conductors can be made accessible will be suitable. Making sure there is no exposed live parts the motor should be run up on load. The ammeter should be clamped round each individual conductor in turn and the three currents recorded. These should be compared with the calculated design current. All enclosures should be returned to their safe condition before leaving.

Now try this - Four

Explain why it is necessary for connections to be shielded to IP2 when voltage tests are to be carried out.

Using Meter Readings

Meter readings are often taken to confirm or clarify a situation. We have seen in question 4 how this could be carried out in one set of circumstances to check design details. It is sometimes necessary to take readings  to sort out a problem or confirm a suspicion. This often requires putting some of the theory into practice. A cable starts to run warm and after tests a piece of equipment is found to be taking more current than it should. The voltage at the terminals of a distribution board is less than it was designed to be.

These are examples of practical situations that need to put theory into practice to come up with a solution to the problem.

Question 5

A bank of fluorescent luminaires appears to be drawing more current than it was designed for. An ammeter reading indicates that a current of 8A is flowing and the circuit voltage is 240V at 50Hz. A wattmeter reading shows the power is 960W. The customer wants to know (a) what the power factor is and (b) what the current would be if the bank of fluorescent luminaires was corrected to unity power factor.

Answer

a. The power factor $= \dfrac{\text{true watts}}{\text{volt amperes}} = \dfrac{960}{240 \times 8} = 0.5$

b. The current, if corrected to unity, can be found from the wattmeter reading as at unity power factor the watts would equal the volt amperes

Current at unity $= \dfrac{\text{power}}{\text{volts}} = \dfrac{960}{240} = 4A$

Now try this - Five

A choke from a fluorescent luminaire is tested on a bench and the following readings were obtained.

> *An a.c. supply of 160V was connected and 8 amperes flowed.*
> *When a d.c. supply of 40V was connected 2.5A flowed.*

It is required to know the power factor of the choke.

Tips to help you answer the "Now try this" questions in this section

Now try this - One

In Part Seven of the IEE Wiring Regulations a check list is given which should be consulted.

Now try this - Two

Section 10 of the On Site Guide may help when looking at safety.

Now try this - Three

The details given in Section 10.3.4 of the On Site Guide should be of help but need modification to apply to the three-phase motor.

Now try this - Four

Look at a copy of the IP code and see what restrictions IP2 offers.

Now try this - Five

On d.c. $\dfrac{U}{I}$ gives the resistance but when the same formula is used for an a.c. circuit the impedance is calculated. Once R and Z have been determined the power factor can be calculated.

Further Revision

1. An inspection reveals that a protection device to BS 88 has been replaced with one to BS 3036. Explain what effect this could have on the circuit in the event of:

    a. an overload
    b. a short circuit

2. Explain the procedure for carrying out an insulation resistance test from a consumer unit before the supply is connected.

3. Describe how an earth fault loop impedance test can be carried out on a single-phase motor circuit.

4. Draw an a.c. sine wave and indicate on it the following:

    a.i)peak value
    ii) rms value

    b. show what is meant by

    i)   amplitude
    ii) frequency

5.a. Explain how to determine the following values on an a.c. single-phase induction motor circuit:

    i)   voltage at the motor
    ii)   current taken by the motor
    iii) power consumed by the motor

    b. Describe the safety precautions that should be applied when the readings in (a) are taken.

Multiple-choice Questions

Write the letter of the correct answer in the box provided in the answer grid at the end of this paper.

1. An employer is required to report an accident to the Health and Safety Executive if an employee is

   a. treated at work for a cut hand
   b. sent home for 24 hours due to the accident
   c. unable to work for more than three days due to the accident
   d. at a hospital casualty department for a morning before returning to work

2. If the electricity supply company states that the nominal supply voltage is 240V, the minimum voltage they can legally supply is

   a. 225.6V
   b. 230.4V
   c. 234V
   d. 240V

3. The supply voltage to a substation is 11kV and has a line current of 100A to delta connected windings. The current flowing in the phase winding of the transformer is

   a. 50A
   b. 57.8A
   c. 100A
   d. 173.2A

4. A high breaking capacity fuse has a British Standard number of

   a. BS3871 Type 2
   b. BS3036
   c. BS1361
   d. BS88 parts 2 & 6

5. A table lamp has a 5A fuse in the plug which is in a 13A socket protected by a 30A mcb. If the consumer unit is protected by a 100A fuse, which is the fuse that should operate when the flex on the lamp shorts out?

   a. 5A
   b. 13A
   c. 30A
   d. 100A

6. A domestic cooker is rated at 27kW when connected to a 240V supply. If diversity is allowed at 10A + 30% full load in excess of the 10A, the supply cable would have an assumed current demand of

   a. 30.75A
   b. 40.75A
   c. 45.75A
   d. 112.5A

7. A single-phase motor is connected so that its voltage, current and wattage can be monitored. One set of readings gives V = 240V, I = 6.25A and P = 1.2kW. The power factor in this case is

   a. 0.5
   b. 0.6
   c. 0.8
   d. 1.25

8. If a supply is said to be a TN-C-S system the earth path from the supply transformer is

   a. combined throughout the supply and consumer's premises
   b. separate throughout the supply and consumer's premises
   c. combined to the consumer's premises then separate
   d. separate to the consumer's premises and then combined

9. A triple pole and neutral isolator consists of

   a. one triple pole and one single pole switch
   b. one triple pole switch and a neutral link
   c. two double pole switches
   d. one four pole switch

10. Which of the following is correct for the bridge rectifier in Diagram MC1?

	A	B	C	D
a.	dc-	a.c.	dc+	a.c.
b.	dc-	dc+	a.c.	a.c.
c.	a.c.	a.c.	dc+	dc-
d.	a.c.	dc+	a.c.	dc-

Diagram MC1

11. If each of the cells shown in Diagram MC2 has an internal resistance of 0.03Ω the total resistance of the battery would be

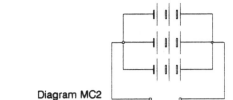

a. 0.01Ω
b. 0.03Ω
c. 0.09Ω
d. 0.27Ω   Diagram MC2

12. The fuse recommended for use in the test probes of a voltmeter should not exceed

a. 30mA
b. 500mA
c. 1A
d. 2.5A

13. A meter is designed to read full scale when 100V is put across it. The turns ratio of a voltage transformer, required so that this meter can be used to monitor the voltage on supply with a maximum of 11kV, will be

a. 9:1
b. 10:1
c. 90:1
d. 110:1

14. A suitable wiring system for a cow shed would be

a. galvanised steel conduit
b. black enamel steel conduit
c. high impact pvc conduit
d. bare mims cable

15. The correction factor of 0.725 is applied when a protection device is installed to

a. BS88 parts 2 & 6
b. BS1361
c. BS3871 Type 2
d. BS3036

16. If a fault develops as indicated in Diagram MC3 the protection device that should operate is

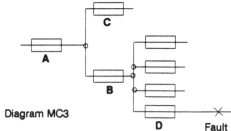

a. A
b. B   Diagram MC3
c. C
d. D

17. The correction factor identified as Ct is for

a. ambient temperature
b. operating temperature of the conductor
c. thermal insulation
d. two groups of conductors

18. In conditions where the supply voltage drops the maximum of 6% and the voltage drop within an installation is 4%, if the nominal declared voltage is 240V the voltage at the load is

a. 240V
b. 230V
c. 225.6V
d. 216.6V

19. A cable has a tabulated voltage drop of 18mV/A/m. If the cable is 25m long, carries a total load current of 17A and is protected by a 20A BS88 fuse the maximum voltage drop at the load is

a. 7.65V
b. 8.50V
c. 9.00V
d. 15.30V

20. A circuit has a load current of 28A and is protected by a 32A BS88 fuse. If correction factors are: for ambient temperature 0.82 and grouping 0.65, the design current for the circuit is

a. 28A
b. 32A
c. 52.5A
d. 60A

21. The minimum permissible cross sectional area for aluminium conductors is

   a. $2.5mm^2$
   b. $4.0mm^2$
   c. $6.0mm^2$
   d. $16.0mm^2$

22. The total earth fault loop impedance path can be calculated from

   a. $Ze = Zs + R1 + R2$
   b. $Zs = Ze + R1 + R2$
   c. $Ze = Zs + R1 - R2$
   d. $Zs = Ze + R1 - R2$

23. A BS 3036 fuse rated at 20A when carrying a fault current of 100A will operate in

   a. 1.8 seconds
   b. 8.5 seconds
   c. 0.85 seconds
   d. 1.5 seconds

24. The current carrying conductor in Diagram MC4 will move

   a. downwards
   b. upwards
   c. to the left
   d. to the right

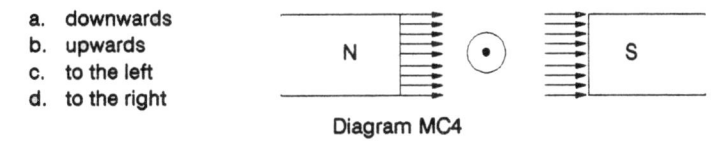

Diagram MC4

25. A three-phase induction motor can have the direction of rotation reversed by reversing

   a. the start windings
   b. the run windings
   c. any two phases
   d. all three phases

26. The synchronous speed of a six pole induction motor operating from a 50Hz supply is

   a. 500 rev/min
   b. 1500 rev/min
   c. 600 rev/min
   d. 1000 rev/min

27. The total current I_T of the circuit in Diagram MC5 is

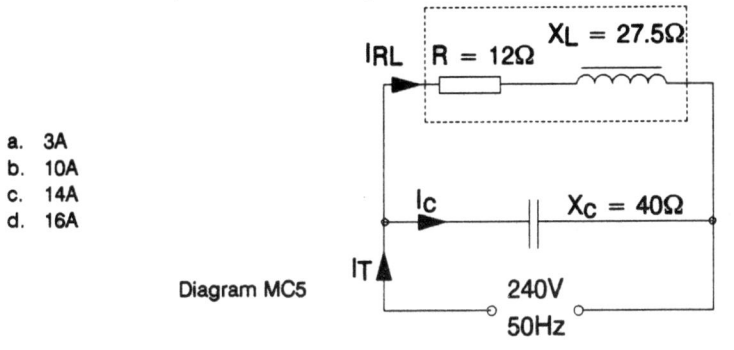

   a. 3A
   b. 10A
   c. 14A
   d. 16A

Diagram MC5

28. A 415V, three-phase motor has an input power of 5kW and a power factor of 0.75 lagging. The line current taken by the motor is

   a. 7.7A
   b. 9.27A
   c. 23A
   d. 16A

29. The purpose of the start winding of a single-phase motor is to

   a. increase the circuit resistance
   b. reduce the starting power
   c. produce a starting torque
   d. reduce the power factor

30. The appropriate type of starter for starting a three-phase 2.75kW cage rotor motor against a very light load is

   a. auto-transformer
   b. rotor resistance
   c. direct-on-line
   d. star delta

31. A motor is stamped with the symbol in Diagram MC6. An application for such a motor could be

   a. cooling electronic equipment
   b. central heating pump
   c. in a freshwater well
   d. petrol pump

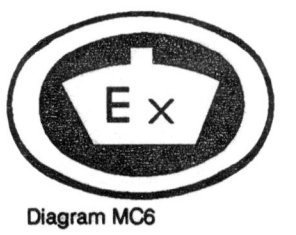

Diagram MC6

32. The power in a three-phase delta connected circuit may be calculated from:

a. $P = \sqrt{3} \ UICos\ \emptyset$

c. $P = UICos\ \emptyset$

b. $P = \dfrac{UI}{\sqrt{3} \ \ Cos\ \emptyset}$

d. $P = \dfrac{\sqrt{3}}{UICos\ \emptyset}$

33. A capacitor start split-phase motor consists of start windings in series with the

a. run windings, centrifugal switch and capacitor
b. centrifugal switch and run windings
c. centrifugal switch and capacitor
d. run windings and capacitor

34. Lumens per watt is the measurement of

a. luminance intensity
b. luminous efficacy
c. luminous flux
d. luminance

35. If X = 24, R = 18 (in Diagram MC7) the value of Z must be

a. 24
b. 30
c. 36
d. 42

Diagram MC7

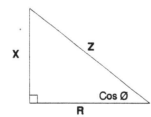

X   Z   Cos Ø   R

36. A capacitor can be connected across the supply to a discharge lamp to

a. reduce the current flowing through the lamp control equipment
b. correct the power factor effect of the lamp and control gear
c. reduce radio interference due to the lamp starting
d. smooth out the supply to the lamp

37. A GLS lamp has an efficacy between

a. 10 to 18 lm/watt
b. 32 to 58 lm/watt
c. 55 to 120 lm/watt
d. 60 to 78 lm/watt

38. Inductive reactance is measured in

a. inductors
b. farads
c. ohms
d. henries

39. For a capacitor to correct the power factor to unity it must have an X_C that equals

a. R
b. Z
c. X_L
d. $\cos\ \emptyset$

40. If the starter switch of a switch start fluorescent fitting is removed when the lamp is working the effect will be that the lamp

a. stays working normally
b. blows the circuit fuse
c. will start flashing
d. goes out

41. The normal operating frequency of a high frequency fluorescent circuit is approximately

a. 50Hz
b. 16kHz
c. 30kHz
d. 50kHz

42. The symbol shown in Diagram MC8 represents a

a. diode
b. triac
c. diac
d. thyristor   Diagram MC8

43. The auxiliary electrode in the discharge tube of a high pressure mercury vapour lamp shown in Diagram MC9 is

a. A
b. B
c. C
d. D

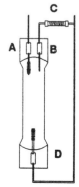

Diagram MC9

44. A continuity test on main equipotential bonding conductors should be carried out with a

a. bell tester
b. milliohmmeter
c. Megohmmeter
d. earth ohmmeter

45. The permitted insulation test voltage on installations supplied with up to 500V a.c. r.m.s. is

a. 250V d.c. at 0.5mA
b. 250V d.c. at 1.0mA
c. 500V d.c. at 0.5mA
d. 500V d.c. at 1.0mA

46. The BS3535 shaver socket is an example of protection by

a. electrical separation
b. barriers and enclosures
c. Class 1 equipment
d. a solid earthed system

47. The resistance between two adjacent heating pipes is found to be 1MΩ. This means that

a. the main equipotential bonding is faulty
b. the two pipes need bonding together
c. a test should be carried out to earth
d. the resistance is satisfactory

48. An enclosure that restricts the access of objects around barriers of 1mm diameter conforms to

a. IP1
b. IP2
c. IP3
d. IP4

49. A polarity test is carried out to confirm that all

a. switches and control devices are in the phase conductors
b. connections are electrically and mechanically sound
c. socket outlets are connected to a ring final circuit
d. circuits will function as intended

50. To test that an RCD will operate under fault conditions satisfactorily, tests are carried out at 50% and 100% of the rated tripping current and 150mA. The trip should operate on

a. only 50%
b. all three tests
c. 50% and 100% only
d. 100% and 150mA only

ANSWER GRID

1. ☐	11. ☐	21. ☐	31. ☐	41. ☐
2. ☐	12. ☐	22. ☐	32. ☐	42. ☐
3. ☐	13. ☐	23. ☐	33. ☐	43. ☐
4. ☐	14. ☐	24. ☐	34. ☐	44. ☐
5. ☐	15. ☐	25. ☐	35. ☐	45. ☐
6. ☐	16. ☐	26. ☐	36. ☐	46. ☐
7. ☐	17. ☐	27. ☐	37. ☐	47. ☐
8. ☐	18. ☐	28. ☐	38. ☐	48. ☐
9. ☐	19. ☐	29. ☐	39. ☐	49. ☐
10. ☐	20. ☐	30. ☐	40. ☐	50. ☐